Trace Elements in Fuel

Trace Elements in Fuel

Suresh P. Babu, *Editor*

A symposium sponsored by
the Division of Fuel
Chemistry at the 166th
Meeting of the American
Chemical Society, Chicago,
Ill., August 30, 1973.

ADVANCES IN CHEMISTRY SERIES **141**

AMERICAN CHEMICAL SOCIETY

WASHINGTON, D. C.   1975

Library of Congress CIP Data

Trace elements in fuel.
  (Advances in chemistry series; 141 ISSN 0065-2393)

  Includes bibliographies and index.

  1. Coal—Analysis—Addresses, essays, lectures.
  2. Trace elements—Addresses, essays, lectures.
  I. Babu, S. P. II. American Chemical Society. Division of Fuel Chemistry. III. Series: Advances in chemistry series; 141.

QD1.A355 no. 141 [TP325]   540'.8s [662'.622]
                                                          75-15522
ISBN 0-8412-0216-8   ADCSAJ 141 1–217

Advances in Chemistry Series

Robert F. Gould, *Editor*

FOREWORD

ADVANCES IN CHEMISTRY SERIES was founded in 1949 by the American Chemical Society as an outlet for symposia and collections of data in special areas of topical interest that could not be accommodated in the Society's journals. It provides a medium for symposia that would otherwise be fragmented, their papers distributed among several journals or not published at all. Papers are refereed critically according to ACS editorial standards and receive the careful attention and processing characteristic of ACS publications. Papers published in ADVANCES IN CHEMISTRY SERIES are original contributions not published elsewhere in whole or major part and include reports of research as well as reviews since symposia may embrace both types of presentation.

CONTENTS

PREFACE

Since the late sixties, social concern about preserving the environment has encouraged the search for methods to convert polluting energy resources into clean-burning fuels. The result of this pursuit has been a major and vital transformation in the energy minerals extraction and utilization methods practiced or abused during the energy "bonanza." Of the alternate energy resources, coal is available in abundance, easy to extract, and the technology is at hand to turn it into pollution-free fuels. As the price of crude oil rises, the processing of coal and oil shale is appearing more and more economically attractive. However, the increased manufacture and distribution of substitute fuels derived from coal should also include cautious monitoring of those elements in the complex coal substance which are potentially harmful to plant and animal life.

The Clean Air Act of 1970 declared beryllium, mercury, and asbestos as hazardous elements. Of the three, mercury is of particular interest to coal technologists. Other elements that exist as trace metals in coal and are suspected to be potentially detrimental to the environment include Pb, As, Sb, Zn, Se, Mo, Co, Li, V, Cr, Mn, Ni, etc. It is not the purpose of this book to create villains out of these elements, but to illustrate analytical techniques to determine how and in what amounts they are released in coal conversion processes.

Systematic efforts to study the origin of minerals in coal did not start until the early or middle thirties. The Environmental Protection Agency is mainly responsible for stimulating research to identify and monitor trace elements in coal processing. After substantial fundamental study, the minerals contained in coal have been classified into syngenetic (inherent in the original plant material) and epigenetic (secondary elements introduced during coalification) elements. Subsequent revelations have shed considerable light on the organic affinity of inorganic elements contained in the mineral matter of coal. These studies have indicated that the syngenetic elements have a greater affinity to organic compounds than the epigenetic elements of coal. It would be worthwhile to follow this basic approach and establish by adequate experimentation (1) the ease of physical separation in coal preparation and (2) the relative volatility under ambient conditions simulating coal conversion processes of the syngenetic and epigenetic elements.

Simultaneously with the efforts to determine the origin of mineral matter in coal, systematic efforts were underway to estimate the quantitative distribution of trace and minor elements in American coals. The early analyses were performed on high-temperature ashes, and as a consequence, the investigators had to be content with determining the nonvolatile metallic oxides. However, with the advent of the low temperature asher and improvisations and advances in wet chemical, radiochemical, and instrumental analytical techniques, we not only can analyze nondecomposed mineral matter but also can study the composition of whole coal.

The contents of this book can be broadly divided into the following categories:

1. Origin and distribution of mineral matter in coal
2. Qualitative and quantitative analysis of trace metals in coal
3. Fate of trace metals in coal preparation and conversion processes
4. The immunopathogenic effects of trace metals released from coal extraction and use

It is my belief that the contents of this volume and the related studies that follow will complement the rational coal utilization route both to produce the necessary energy and to protect the plant and animal life—a goal towards which we direct our efforts so earnestly.

Institute of Gas Technology   Suresh P. Babu
Chicago, Ill.
August 1974

Mineral Matter and Trace Elements in Coal

HAROLD J. GLUSKOTER

Illinois State Geological Survey, Urbana, Ill. 61801

The term, "mineral matter in coal," refers to mineral phases or species present in coal and also to all chemical elements in coal that are generally considered to be inorganic. Most mineral matter occurs in coal as silicates, sulfides, and carbonates. Four coals, separated into series of specific gravity fractions, have been analyzed. The trace elements, germanium, beryllium, and boron, have the greatest organic affinities, whereas Hg, Zr, Zn, As, Cd, Pb, Mn, and Mo are generally inorganically combined in the coal. Each of the other trace elements determined apparently occurs in both organic and inorganic combination. P, Ga, Sb, Ti, and V are more closely associated with the elements having strong organic affinities while Co, Ni, Se, Cr, and Cu are more closely associated with the elements having strong inorganic affinities.

The term, "mineral matter in coal," is widely used, but its meaning varies appreciably. The term usually includes all inorganic non-coal material found in coal as mineral phases and also all elements in coal that are considered inorganic. Therefore, all elements in coal except carbon, hydrogen, oxygen, nitrogen, and sulfur are included in this broad definition. Four of these five organic elements also are found in coals in inorganic combination and therefore are part of the mineral matter. Carbon is present in carbonates [$Ca(Fe,Mg)CO_3$]; hydrogen in free water and water of hydration; oxygen in water, oxides, carbonates, sulfates, and silicates; and sulfur in sulfides (primarily pyrite and marcasite) and sulfates.

Interest in mineral matter in coal arises primarily because some of the materials may have detrimental effects during coal use. Because methods of using coal are becoming more sophisticated and increasingly large amounts of coal are being used at single locations, these detrimental effects should be considered further.

Although the amount of mineral matter in coals varies considerably, it is normally large enough to be significant however the coal is used. In a study of 65 Illinois coals, Rao and Gluskoter (1) found the mineral matter content to range from 9.4 to 22.3%, corresponding to an ash content of 7.3 and 15.8%, respectively. O'Gorman and Walker (2) found an even larger range (9.05–32.26%) in mineral matter content in 16 whole coal samples from a wide distribution of locations in North America. If 15% is a reasonable estimate of the average value of mineral matter content in coals mined in North America, then, unless partially removed by cleaning, that amount enters each coal utilization process. Since approximately 590×10^6 tons of coal were produced in the United States in 1973 it is estimated that 89×10^6 tons of this was normally unwanted mineral matter.

Interest in mineral matter content of coal intensified as electric power plants became larger and the boilers began to operate at higher temperatures. Problems of fireside boiler-tube fouling and corrosion, which became increasingly severe at the higher temperatures, were related to the sulfur, chlorine, alkali, and ash content of the coals (3). Within the past several years, the general public has become more interested in both air and water pollution. Therefore, both the coal consumer and the producer need a more thorough knowledge of the mineral matter in coal and of the products and by-products of the mineral matter produced when coal is combusted. Much of this interest has been directed to the forms of sulfur in coal and coal refuse, to the sulfur oxides formed during coal combustion, and to the sulfates from coal oxidation. There has been a consequent demand for data relating to the origin, distribution, and reactions of sulfur in coal. There is also much demand for data on the trace elements in coals: their concentrations and distributions in coals, their volatility, and their potential effects on the environment.

More recently, there has been much concern about the possible effects of the mineral matter in coal on processes used to convert coal to other fuels such as gasification, liquefaction, and production of clean solid fuels. Not only is removing and disposing of the mineral matter a problem, but also the possible chemical effects such as catalyst poisoning, which might be expected in the methanation of gas from coal, should be considered.

Not all of the interest in mineral matter in coals is stimulated by its detrimental effects during coal use. In several instances coal is a source of desired elements and materials. Uranium has been produced from lignite; germanium and sulfur could be produced from coal; and coal ash has been used for construction materials such as brick, lightweight aggregate, and road paving material.

Minerals in Coal

Several dozen minerals are reported in coals, although most of these occur only sporadically or in trace amounts. The overwhelming majority of the minerals in coal are in one of four groups: aluminosilicates, carbonates, sulfides, and silica (quartz) (Figure 1).

Aluminosilicates—Clay Minerals. Clay minerals are the most commonly occurring inorganic constituents of coals and of the strata associated with the coals. Much of the work on clays reported in the literature is concerned with these strata, not the coals themselves. Many different clay minerals have been reported in coals, but the most common are illite $[(OH)_4K_2(Si_6 \cdot Al_2)Al_4O_{20}]$, kaolinite $[(OH)_8Si_4Al_4O_{10}]$, and mixed-layer illite–montmorillonite. Rao and Gluskoter (*1*), in an investigation of 65 coals from the Illinois Basin, reported a mean value of 52% for clay in the mineral matter. O'Gorman and Walker (*2*) also found that the clay minerals make up the greater part of the mineral matter in most of the coals that they studied.

Sulfides and Sulfates. Pyrite is the dominant sulfide mineral in coal. Marcasite has also been reported from many different coals. Pyrite and marcasite are dimorphs, minerals that are identical in chemical composition (FeS_2) but differ in crystalline form; pyrite is cubic while marcasite is orthorhombic. Other sulfide minerals that have been found in coals, and sometimes in significant amounts, are sphalerite (ZnS) and galena (PbS).

Sulfates are not common and often are not present at all in coals that are fresh and unweathered. Pyrite is very susceptible to oxidation and decomposes to various phases of iron sulfate minerals at room temperature. The following iron sulfate mineral phases are associated with Illinois coals that have been subjected to oxidizing conditions (*4*):

szomolnokite	$FeSO_4 \cdot H_2O$
rozenite	$FeSO_4 \cdot 4H_2O$
melanterite	$FeSO_4 \cdot 7H_2O$
coquimbite	$Fe_2(SO_4)_3 \cdot 9H_2O$
roemerite	$FeSO_4 \cdot Fe_2(SO_4)_3 \cdot 12H_2O$
jarosite	usually a sodium jarosite $(Na,K)Fe_3(SO_4)_2(OH)_6$

Sulfides and sulfate minerals make up 25% of the mineral matter content of Illinois coals (*1*).

Carbonates. The carbonate minerals, in general, vary widely in composition because of the extensive solid solution of calcium, magnesium, iron, manganese, etc. that is possible within them. There is also a wide range of mineral compositions for the carbonate minerals in coals. The relatively pure end members, calcite ($CaCO_3$) and siderite ($FeCO_3$),

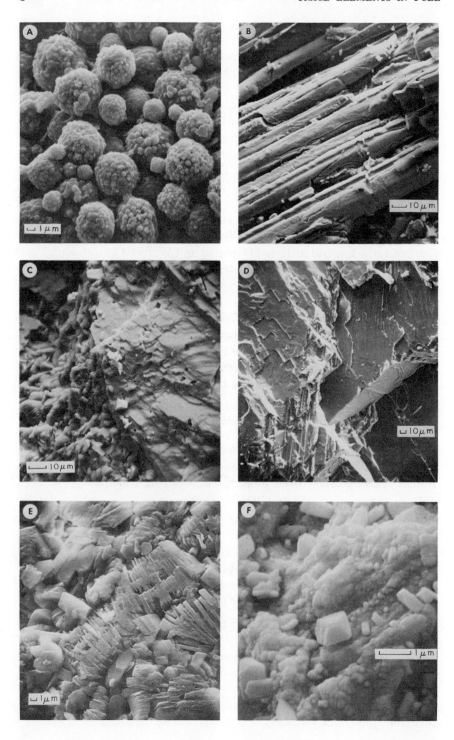

have commonly been reported. However the most frequently reported carbonate minerals from the majority of the coals in the world are dolomite ($CaCO_3 \cdot MgCO_3$) and ankerite ($2CaCO_3 \cdot MgCO_3 \cdot FeCO_3$).

There are significant differences in the carbonate mineralogy of coals from different parts of the world. Calcite is nearly the only carbonate mineral observed in Illinois coals (1) whereas ankerite is the dominant carbonate mineral in British coals (5, 6), and siderite, ankerite, and calcite are common in Australian coals (7). The carbonate minerals make up 9% of the total mineral matter content of coals from the Illinois Basin (1).

Silica (Quartz). Quartz is ubiquitous in all coals. Rao and Gluskoter (1) reported that, on the average, 15% of the mineral matter in coals from the Illinois Basin was quartz. O'Gorman and Walker (2) found 1–20% quartz in 16 whole coal samples from various parts of the United States.

Mineral Matter in Coal and High-Temperature Coal Ash

The mineral matter content of coal cannot be determined qualitatively or quantitatively from the ash that is formed when the coal is oxidized. Normal high-temperature ashing of coal at 750°C, as designated by ASTM standards (8), causes a series of reactions involving the minerals in the coal. Of the four major mineral groups only quartz is not altered during high-temperature ashing.

Clay Minerals. The clay minerals in coal all contain water bound within their lattices. Kaolinite contains 13.96%, illite 4.5%, and montmorillonite 5% bound water. In addition, the montmorillonite in the mixed-layer clays also contains interlayer or adsorbed water. All of the water is lost during the high-temperature ashing.

Iron Sulfide Minerals. During high-temperature ashing, the pyrite minerals are oxidized to ferric oxide and sulfur dioxide. Some of the sulfur dioxide may remain combined with calcium in the ash, but much is lost. If all the sulfur dioxide were emitted during ashing, there would be a 33% weight loss with respect to the weight of pyrite or marcasite in the original sample.

Figure 1. Scanning electron photomicrographs of minerals from coals. The minerals were studied and photographed by a Cambridge Stereoscan microscope with an accessory energy-dispersive x-ray spectrometer at the Center for Electron Microscopy, University of Illinois. A. Pyrite framboids from the low-temperature ash of a sample from the DeKoven Coal Member. B. Pyrite: cast of plant cells from the low-temperature ash of a sample from the Colchester (No. 2) Coal Member. C. Kaolinite (left) and sphalerite (right) in minerals from a cleat (vertical fracture), Herrin (No. 6) Coal Member. D. Calcite from a cleat in the Herrin (No. 6) Coal Member. E. Kaolinite "books" from a cleat in the Herrin (No. 6) Coal Member. F. Galena: small crystals in the low-temperature ash of a sample from the DeKoven Coal Member.

Calcite. The calcium carbonate is calcined to lime (CaO) during high-temperature ashing, with a loss of carbon dioxide. This results in a 44% weight loss.

Quartz. The stable mineral quartz (SiO_2) is the only major mineral found in coal which is inert during high-temperature ashing.

The changes in the mineral matter content in coal during ashing have long been recognized. A number of workers have suggested schemes for calculating the true mineral matter content from determinations made during the chemical analyses of coal. Parr (9) reported one of the earliest of such schemes in which only the total sulfur and ash contents were considered in developing the conversion formulae. This is still the most widely used procedure. A more sophisticated method was suggested by King *et al.* (10), which considers both the carbon dioxide loss from carbonates and the chloride decomposition in addition to the factors considered by the Parr formula. Further modifications of these techniques have been suggested by Brown *et al.* (11), Pringle and Bradburn (5), and Millot (12). Recently Given (13) discussed in detail the problem of converting raw chemical analytical data to a pure coal basis by subtracting the calculated mineral matter content.

Analyses of Minerals in Coal

Separation of Minerals from Coal. This section concerns minerals in the strict sense—naturally occurring, inorganic substances with definite chemical composition and ordered atomic arrangement—and not individual chemical elements. There have been many studies in which workers have analyzed minerals that were picked by hand from coal seams, from coal partings, or from cleats or that were separated from the coal by a method based on differences in specific gravity between the coal and the minerals contained in it. As a first step, these studies were important contributions, but it quickly became apparent that the large amount of inseparable mineral matter in coals resulted in incomplete analytical data. In the next stage in this research the coals were ashed at 300°–500°C, temperatures below that of normal combustion, or at room temperatures in an oxygen stream. This provided additional information, but was limited because many of the minerals were oxidized along with the organic fraction.

Within the past decade the technique of electronic (radiofrequency) low-temperature ashing has been used to investigate mineral matter in coal. In a low-temperature asher, oxygen is passed through a radiofrequency field, and a discharge takes place. Activated oxygen thus formed passes over the coal sample, and the organic matter is oxidized at relatively low temperatures—usually less than 150°C (14).

The effects of low-temperature ashing and of the oxidizing gas stream on the minerals in coal are minimal. No oxidation of mineral phases present has been reported, and the only phase changes observed were those expected at 150°C and 1 torr. Therefore, most of the major mineral constituents of coals, including pyrite, kaolinite, illite, quartz, and calcite, are unaffected by the radiofrequency ashing. Recent studies of mineral matter in coal which used radiofrequency low-temperature ashing include Gluskoter (15), Estep *et al.* (16), Wolfe (17), O'Gorman and Walker (2), and Rao and Gluskoter (1).

Identification of Minerals in Coal. Once the low-temperature mineral matter residue has been obtained by radiofrequency ashing, the minerals can be identified, and their concentrations can be determined by a variety of instrumental techniques. The best developed, most inclusive, and probably most reliable method used thus far in distinguishing minerals in coal is x-ray diffraction analysis. It has been used extensively by Gluskoter (15), Wolfe (17), O'Gorman and Walker (2), and Rao and Gluskoter (1) and has been somewhat successful in quantifying mineral analyses.

Estep *et al.* (16) used infrared absorption bands in the region 650–200 cm^{-1} to analyze quantitatively as well as qualitatively for minerals in low-temperature ash. O'Gorman and Walker (2) also applied this technique in their investigations.

Differential thermal analyses (DTA) of minerals in a high-temperature coal ash have been reported by Warne (18, 19). The method was applied to the mineral matter fraction of four samples by O'Gorman and Walker (2).

Electron microscopy, even though it has rapidly increased in popularity as a mineralogical research tool, has not been used extensively to identify minerals in coals. Dutcher *et al.* (20) reported on a limited investigation which used the electron probe to analyze mineral matter in coal. Scanning electron microscopy with an energy-dispersive x-ray system accessory has been used to a limited extent to study minerals obtained from the low-temperature ashing of coal (21, 22, 23). A study of ^{57}Fe Mössbauer spectra in coals by Lefelhocz *et al.* (24) demonstrated the validity of applying this technique to coal and suggested the presence of high-spin iron(II) in six-fold coordination in several of the samples studied.

Mass spectrometric investigations of isotopes in coal and coal minerals have also been very limited in scope. Rafter (25) published sulfur isotope data on 27 New Zealand coal samples but did not draw any conclusions from these data. Smith and Batts (26) determined the isotopic composition of sulfur in a number of Australian coals and concluded that, from this type of data, one might deduce the origin of the organically combined sulfur, the depth of penetration of sea water into underlying coal measures, and the factors controlling reduction of sulfates to sulfides by biogenic residues.

Investigation of Minerals in Coal *in Situ*. The coal petrographer uses the optical microscope, usually in reflected light mode, to characterize the organic fraction (macerals) in coals. The science of coal petrography has developed a high degree of precision, particularly during the last 25 yrs. However, these techniques have not been nearly as successful in investigating mineral matter in coal. Pyrite, because of its high reflectance and its abundance in coals, is the most likely mineral to be studied microscopically. Several automated microscopes and image-analyzing microscopes have been developed that could be used in such a study. McCartney and Ergun (*27*) reported on an automated reflectance scanning microscope system of their own design and included results of pyrite analyses.

Geochemistry of Mineral Matter in Coal

As previously discussed the major problem of obtaining an unaltered mineral-matter residue has been alleviated somewhat by using radiofrequency low-temperature ashing. Another serious problem, and one less amenable to solution, is the complexity of this system of mineral matter in coal. The complexity results from the variety of physical and chemical conditions in which the coal-forming materials were deposited and in which the coal formed. The system of mineral matter in coal is a relatively low-temperature, low-pressure system with many component phases. It is an open system with many mobile components. There is also the further complication that the system has been active and may have been changing at any time since its genesis (approximately 300 million yrs for coals of the Pennsylvanian system). Although the system is complicated, interpreting it should not be impossible, for the minerals associated with coal and with all other sedimentary rocks are not the results of random deposition. They are the predictable end product of a definite set of biological, chemical, and physical conditions, which provided an environment in which the minerals could be deposited or in which they could form.

The mineral matter and the ash in coal have often been informally classified as inherent (stemming from the plant material in the coal swamp) or as adventitious (added after the deposition of the plant material in the swamp). This classification is misleading and difficult to apply, especially for those minerals that are contemporaneous with the peat swamp but were not incorporated by the plants.

There are standard terms applied to sediments and sedimentary rocks that can be used with coal minerals. Those minerals which were transported by water or wind and deposited in the coal swamp are allogenic or detrital. All of the minerals which formed within the coal swamp, in the peat, or in the coal are authigenic. The term syngenetic applies to the minerals that were contemporaneous with the coal formation, and epigenetic refers to those which were formed later, such as cleat fillings.

Chemical Analyses of Mineral Matter and Trace Elements in Coal

Inasmuch as mineral matter has been defined broadly to include all inorganic elements in coals, the chemical characterization of mineral matter involves the determination of many elements. In general, chemical analyses of geological materials have progressed from the wet chemical methods to sophisticated instrumental methods. The major elements in the mineral constituents of coal, Si, Al, Ti, Ca, Mg, Fe, P, S, Na, K, are the same as those in silicate rocks and are often determined by x-ray fluorescence spectroscopy and flame photometry.

The minor and trace elements in coals are currently determined by several techniques, the most popular of which are optical emission and atomic absorption spectroscopy. Neutron activation analysis is also an excellent technique for determining many elements, but it requires a neutron source, usually an atomic reactor. In addition, x-ray fluorescence spectroscopy, electron spectroscopy for chemical analyses (ESCA), and spark source mass spectroscopy have been successfully applied to the analyses of some minor and trace elements in coal.

Until recently, chemical analyses of coals were done on ash produced from the coal at relatively high temperatures. This was the standard approach for many years, and analyses of trace elements in coals do have a long history. An early article on an element as rare as cadmium in coal was published 125 yrs ago (28). One limitation of high-temperature ash sample is that volatile elements may be lost during combustion and will not be detected. Another problem which applies especially to analyses for trace and minor elements is that there have not been any coal standards available until very recently.

Recent comprehensive investigations involving a large number of coal samples and determinations of many elements including trace elements have been undertaken by the U.S. Geological Survey (29), the U.S. Bureau of Mines (30), the Illinois State Geological Survey (23), and The Pennsylvania State University (2).

Literature of Trace Elements and Mineral Matter in Coal

A detailed review of the world literature concerned with mineral matter and trace elements in coal is well beyond the scope of this chapter. Such a review would involve the discussion of several thousand books, journal articles, and other publications and would itself compose at least a modest volume. However, the annotated bibliography at the end of this chapter lists several review articles that may be of help to anyone interested in obtaining further information on trace elements and mineral matter in coal.

Occurrence of Trace Elements in Coal

The modern investigations of trace elements in coals were pioneered by Goldschmidt, who developed the technique of quantitative chemical analysis by optical emission spectroscopy and applied it to coal ash. In these earliest works, Goldschmidt (31) was concerned with the chemical combinations of the trace elements in coals. In addition to identifying trace elements in inorganic combinations with the minerals in coal, he postulated the presence of metal organic complexes and attributed the observed concentrations of vanadium, molybdenum, and nickel to the presence of such complexes in coal.

Goldschmidt (32) also introduced the concept of a geochemical classification of elements, in which the elements are classified on the basis of their affinities and tendencies to occur in minerals of a single group. The chalcophile elements are those which commonly form sulfides. In addition to sulfur, they include Zn, Cd, Hg, Cu, Pb, As, Sb, Se, and others. When present in coals, these elements would be expected to occur, at least in part, in sulfide minerals. Sulfides other than pyrite and marcasite have been noted in coals, but, except in areas of local concentration, they occur in trace or minor amounts.

The lithophile elements are those that generally occur in silicate phases and include among others: Si, Al, Ti, K, Na, Zr, Be, and Y. These would be expected to occur in coals in some combination with the silicate minerals: kaolinite, illite, other clay minerals, quartz, and stable heavy detrital minerals.

The carbonate minerals in coals occur primarily as epigenetic fracture fillings (cleat filling). Magnesium, iron, and manganese are often associated with the sedimentary carbonate minerals and would reasonably be expected to be associated with the cleat fillings in coal.

A large number of silicate, sulfide, and carbonate minerals have been identified from coal seams, and the elements composing them necessarily occur in coals in inorganic combination. However, mineralogical investigations of coals have not generally been quantitative, and whether an element occurs only in inorganic combination or perhaps is also present in organic combination has not commonly been considered.

Nicholls (33) approached this problem by plotting the analytical data for the concentration of a single element in coal or in coal ash against the ash content of the coal. Diagrams depicting a number of such points for a single coal seam or for a group of coal seams in a single geographic area were interpreted for degree of inorganic or organic affinity of the element. Nicholls concluded (33):

. . . one element, boron, is largely, almost entirely, associated with the organic fraction in coals; some elements, such as barium, chromium, cobalt, lead, strontium, and vanadium are, in the majority of cases, associated with the inorganic fraction; and a third group including nickel, gallium, germanium, molybdenum, and copper, may be associated with either of both fractions.

He then subdivided the third group into nickel and copper, which are in inorganic combination when found in large concentrations, and gallium, germanium, and molybdenum, which are largely in organic combination when found strongly concentrated.

Horton and Aubrey (34) handpicked pure vitrain samples from coals and separated them into five different specific gravity fractions. They then analyzed these for 16 minor elements. They concluded that for the three vitrains they studied, beryllium, germanium, vanadium, titanium, and boron were contributed almost entirely by the inherent (organically combined) mineral matter and that manganese, phosphorus, and tin were associated with the adventitious (inorganically combined) mineral matter.

A much more ambitious series of investigations of the organic–inorganic affinities of trace metals in coals was undertaken and reported on by Zubovic and co-workers at the U.S. Geological Survey (35, 36, 37, 38, 39). In the most recent article, Zubovic (39) listed the following 15 elements in order of percent organic affinity: Ge (87), Be (82), Ga (79), Ti (78), B (77), V (76), Ni (59), Cr (55), Co (53), Y (53), Mo (40), Cu (34), Sn (27), La (3), and Zn (0). He concluded that this series was apparently related to the chelating properties of the metals.

The Illinois State Geological Survey has recently been extensively investigating trace elements in coal (23, 40). As a part of this study four sets of float–sink samples were analyzed for a number of trace and minor elements. Three coals, crushed and sized to ⅜ in. by 28 mesh, were separated into six specific gravity fractions by floating them in mixtures of perchloroethylene and naphtha. The heaviest of these six fractions (1.60 sink) was then separated into two parts using bromoform (specific gravity 2.89). The fourth coal was also separated in perchloroethylene and naphtha, but only two fractions were analyzed, one with specific gravity of less than 1.25 and one with specific gravity heavier than 1.60. By use of a technique similar to that of Zubovic (39), the trace elements determined in these samples are listed in order of decreasing affinity for the clean coal fractions, or decreasing organic affinity (Table I). The sequence was determined by comparing ratios of the amount of an element in the lightest float fraction (always less than 1.30 specific gravity) to the amount of the element in the 1.60 sink fraction. The numerical values thus determined are not given because they vary with the particle size distribution of the coal, the specific gravity of the liquid used to make the first (lightest) separation, and the size distribution of the mineral fragments in a single coal. However, the sequence given in Table I does indicate which elements are primarily in organic combination, which are in inorganic combination, and which are, apparently, both inorganically and organically combined in coals of the Illinois Basin.

Table I.   Affinity of Elements for Pure Coal and Mineral Matter as Determined from Float–Sink Data

	Davis Coal	DeKoven Coal	Colchester (No. 2) Coal	Herrin (No. 6) Coal
Clean coal—lightest	B	Ge	Ge	Ge
specific gravity fraction	Ge	Ga	B	B
(elements in "organic	Be	Be	P	Be
combination")	Ti	Ti	Be	Sb
↑	Ga	Sb	Sb	V
	P	Co	Ti	Mo
	V	P	Co	Ga
	Cr	Ni	Se	P
	Sb	Cu	Ga	Se
	Se	Se	V	Ni
	Co	Cr	Ni	Cr
	Cu	Mn	Pb	Co
	Ni	Zn	Cu	Cu
	Mn	Zr	Hg	Ti
	Zr	V	Zr	Zr
	Mo	Mo	Cr	Pb
↓	Cd	Pb	Mn	Mn
Mineral matter—specific	Hg	Hg	As	As
gravity greater than 1.60	Pb	As	Mo	Cd
(elements in "inorganic	Zn		Cd	Zn
combination")	As		Zn	Hg

The sequences shown in Table I can be divided into several general groups. First, there are those elements which are always in the group most closely associated with the clean coal and which, therefore, have the greatest organic affinities. These are germanium, beryllium, and boron, which are three of the top five elements listed by Zubovic (39). At the other end of the list are the elements with the least affinity for the organic portion of the coal. The elements mercury, zirconium, zinc, arsenic, and cadmium are near the bottom in all four coals studied, and lead, manganese, and molybdenum are near the bottom in three of the four. The remaining elements, those that are apparently associated to varying degrees with both the organic and inorganic portions of the coals, can also be divided into two groups: those elements that tend to be more generally allied to the elements with organic affinities (phosphorus, gallium, antimony, titanium, and vanadium) and those elements that tend to be more inorganically associated (cobalt, nickel, chromium, selenium, and copper). This summarized sequence generally agrees with that given by Zubovic (*39*) with only a few minor discrepancies. The elements listed in Table I include 12 of the 15 elements discussed by Zubovic (*39*) as well as nine additional elements.

Although an element is listed among those with the highest organic affinities, its occurrence in inorganic combination in coals is not precluded. Boron, which is among those found in high concentrations in the cleanest coal fractions, occurs in amounts up to 200 ppm in the clay mineral illite from Illinois coals (*41*). Similarly, a portion of those elements usually concentrated most heavily in the high specific gravity fractions may also be in organic combination. This dual occurrence was postulated for the mercury content of Illinois coals by Ruch *et al.* (*42*), and mercury is included here with the elements having the lower organic affinities.

The float–sink, or washability, data can be displayed as washability curves and as histograms. Washability curves and histograms for a series of elements are given in Figures 2–5. The figures are presented in order of the increasing tendencies of the elements to be concentrated in the heavier fractions (decreasing organic affinity). The washability curve is a type of cumulative curve from which one can read the expected concentration of an element at any given recovery rate of a coal, assuming separation based on specific gravity differences. Therefore, the abscissa is "recovery of float coal—percent" and should be applicable to any specific gravity separation without regard to the medium in which it is done or the method used. The raw coal concentration of an element is read at the 100% recovery point, and concentration in the cleanest coals (most mineral matter-free) is read at the low recovery end of the curve (20–30% recovery).

Figure 2 shows the washability curve and histogram for germanium in a sample from the Davis Coal Member. Germanium is the element with the highest organic affinity in the coals studied. The negative slope of the curve indicates that germanium is concentrated in the clean coal fractions; this is also apparent from the histogram. The histogram indicates that there is a higher concentration of germanium in the 1.60–2.79 specific gravity fraction than in the greater-than-2.79 specific gravity fraction. Apparently, a greater portion of the germanium is concentrated with the clay minerals than with the sulfide minerals, which compose the majority of the 2.79 sink fraction.

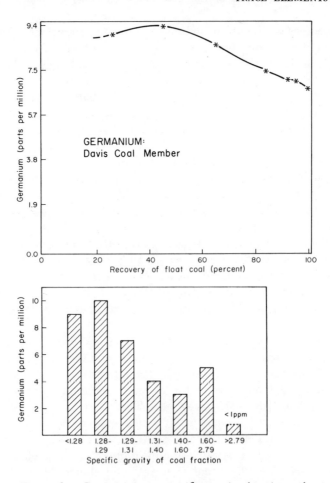

Figure 2. Germanium in specific gravity fractions of a sample from the Davis Coal Member. Upper: washability curve. Lower: distribution of germanium in individual fractions.

All of the washability curves shown are drawn with the ordinates the same length and the origin at zero concentration so that the slopes can be compared and interpreted. Figure 3, which shows beryllium in the Davis Coal Member, presents a flat washability curve and also a relatively uniform histogram. Beryllium is, therefore, rather evenly distributed in the clean coal samples and is also present, in somewhat lesser amounts, in the heavier sink fractions.

Nickel in a sample from the Colchester (No. 2) Coal Member (Figure 4) is definitely concentrated in the heavier specific gravity fractions. However, the washability curve remains well above the abscissa and does not appear to approach the origin in the cleanest fraction (purest

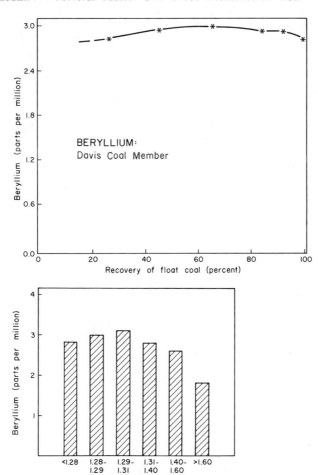

Figure 3. Beryllium in specific gravity fractions of a sample from the Davis Coal Member. Upper: washability curve. Lower: distribution of beryllium in individual fractions.

coal). This pattern may be interpreted as showing nickel in organic as well as in inorganic combination.

The final washability curve and histogram (Figure 5) are of arsenic in a sample from the Herrin (No. 6) Coal Member. The washability curve shows an intense concentration of arsenic in the heavier fractions. The curve approaches the abscissa and if extrapolated would intersect the ordinate near the origin. Arsenic is one of the least organically related elements in all four of the coals studied.

Figures 2–5 present examples of graphic representations of the organic–inorganic affinities of several elements in coals. A larger number

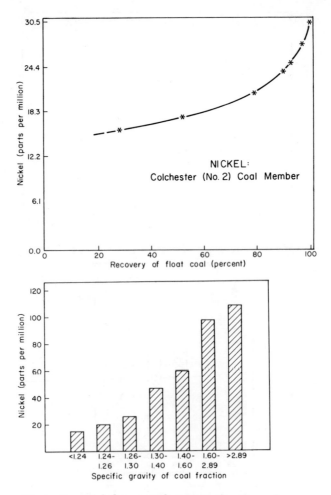

Figure 4. *Nickel in specific gravity fractions of a sample from the Colchester (No. 2) Coal Member. Upper: washability curve. Lower: distribution of nickel in individual fractions.*

of such curves for the elemental distribution of these coals and statistical summaries of major, minor, and trace element distributions in 101 coals from the United States are given in Ruch *et al.* (*40*).

Conclusions

Mineral matter in coal, as the term is generally used, includes the mineral phases (species) present in the coal seam as well as those elements generally thought to be inorganic, even if they are present in coals in organic combination. The major minerals found in coals are silicates

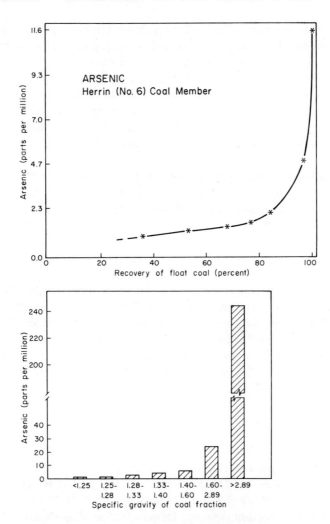

Figure 5. Arsenic in specific gravity fractions of a sample from the Herrin (No. 6) Coal Member. Upper: washability curve. Lower: distribution of arsenic in individual fractions.

(kaolinite, illite, mixed-layer clay minerals, and quartz), sulfides (primarily pyrite and marcasites), and carbonates (calcite, siderite, and ankerite, or ferroan dolomite).

The two major problems encountered in investigating mineral matter in coal are that the mineral matter and coal are so intimately mixed that physical separation of the two is not feasible in a quantitative study and that the geochemical system is extremely complicated and difficult to characterize. The first problem is overcome in part by refining the techniques used to prepare samples for analyses (radiofrequency low-temperature ashing, for example), but the second problem is an inherent part of the system and is less amenable to solution.

The literature concerned with trace elements and minerals in coals has increased in quantity as the interest in these materials, caused by concern about environmental quality and energy availability, has increased. A number of excellent review articles, particularly that by Watt (43), have been published and provide comprehensive summaries of the existing literature.

Four Illinois coals, separated into series of specific gravity fractions, have been analyzed for 21 trace and minor elements. The degree to which an element is associated with the lightest specific gravity fraction and therefore with the purest coal is a measure of that element's organic affinity in coal. If the element is concentrated in the heaviest specific gravity fraction, it is then in inorganic combination. Germanium, beryllium, and boron have been determined to have the greatest organic affinities, and Hg, Zr, Zn, As, Cd, Pb, Mn, and Mo are generally inorganically combined. The other elements that were determined all show degrees of inorganic and organic associations, with P, Ga, Sb, Ti, and V tending to be allied with the other elements having organic affinities and Co, Ni, Cr, Se, and Cu more closely associated with the inorganically combined elements. This series is very similar to a series of elements in coals presented in order of decreasing organic affinities by Zubovic (39).

Acknowledgments

The chemical analytical data on which Table I and Figures 2–5 are based were determined by the Analytical Chemistry Section of the Illinois State Geological Survey. The Survey research reported is sponsored, in part, by Grant No. R-800059 and Contract No. 68-02-0246 from the U. S. Environmental Protection Agency, Demonstration Projects Branch, Control Systems Laboratory, Research Triangle Park, N. C.

Annotated Bibliography

1. Watt, J. D., "The Physical, and Chemical Behaviour of the Mineral Matter in Coal Under Conditions Met in Combustion Plant. Part I, The Occurrence, Origin, Identity, Distribution, and Estimation of the Mineral Species in British Coals," British Coal Utilization Research Association, Literature Survey, 121 p., Leatherhead, Surrey, England, 1968.
    This excellent review of minerals in coal and the chemical composition of coals does not, as the title may suggest, limit itself to British coals. The literature of the rest of Europe and of North America is generously cited. A large section devoted to the various methods of determining the amount of mineral matter is included. There is also a section concerned with the methods of identifying mineral species in coals.

2. Nicholls, G. D., "The Geochemistry of Coal-bearing Strata" in "Coal and Coal-bearing Strata" (D. G. Murchison and T. S. Westoll, Eds.), p. 269–307, Oliver and Boyd, Edinburgh and London, 1968.

Many data on trace elements in coal are included in the chapter. The emphasis is on the geochemistry of the trace elements, with discussions of the concentration levels of trace elements in coals, the organic (or inorganic) affinities of the trace elements, and the geochemical controls of associations of elements. This and the following significant review article are contained within a single volume.

3. Mackowsky, M. Th., "Mineral Matter in Coal" in "Coal and Coal-bearing Strata" (D. G. Murchison and T. S. Westoll, Eds.), p. 309–321, Oliver and Boyd, Edinburgh and London, 1968.

Mackowsky discusses the mineral phases found in coals and differentiates those phases with geologically different genetic histories. A number of the mineral phases are shown in a series of photomicrographs of polished sections of coals.

4. Williams, F. A., Cawley, C. M., "Impurities in Coal and Petroleum" in "The Mechanism of Corrosion by Fuel Impurities" (H. R. Johnson and D. J. Littler, Eds.), p. 24–67, Butterworths, London, 1963.

In addition to discussing the major minerals and the elemental composition of coals, the authors discuss methods of removing impurities from coal and the effects of the impurities on several aspects of coal utilization.

5. Magee, E. M., Hall, H. J., Varga, Jr., G. M., "Potential Pollutants in Fossil Fuels," Environmental Protection Technology Series, EPA-R2-73-249, 293 p., U.S. Environmental Protection Agency, 1973.

The data on sulfur, nitrogen, and trace elements in coal and in oil are summarized. Emphasis is placed on the geographic and geologic distribution of the trace elements.

6. Abernethy, R. F., Gibson, F. H., "Rare Elements in Coal," Information Circular 8163, 69 p., U.S. Bureau of Mines, 1963.

This circular reviews the occurrence of 34 trace elements in coal and the occurrence and distribution of chlorine, phosphorus, titanium, and manganese, which are not considered rare in coal. Separate chapters are devoted to germanium, gallium, and uranium in coal.

7. Averitt, Paul, Breger, I. A., Swanson, V. E., Zubovic, Peter, Gluskoter, H. J., "Minor Elements in Coal—A Selected Bibliography, July 1972, Professional Paper 800-D, p. D-169–D-171, U.S. Geological Survey, 1972.

This bibliography lists 65 selected references to articles on trace elements in coals. Most articles cited are in English and are concerned with North American coals.

8. Swaine, D. J., "Trace elements in Coal" (originally written in English). Published in Russian as Svein, D. Dzh., "Mikroelementy v ugliakh" in "Ocherki sovremonnoĭ geokhimii i analiticheskoĭ khimii" (Symposium volume in honor of the 75th anniversary of Academician A. P. Vinogradov) (A. I. Tugarinov, Ed.), p. 482–492, Moskva: "Nauka" Publishers, 1972.

This review contains sections on the occurrence of 13 trace elements in coals and a summary of trace element distribution in coals.

9. Thiessen, Gilbert, "Composition and Origin of the Mineral Matter in Coal" in "Chemistry of Coal Utilization" (H. H. Lowry, Ed.), v. 1, p. 485–495, John Wiley and Sons, Inc., New York, 1945.

The relationships of mineral matter to ash content are discussed. Concentrations of chemical elements are mentioned, and the mineral matter contents of petrographic components are summarized.

10. Ode, W. H., "Coal Analysis and Mineral Matter" in "Chemistry of Coal Utlization, Supplementary Volume" (H. H. Lowry, Ed.), p. 202–231, John

Wiley and Sons, Inc., New York, 1963.
    Methods of coal analysis are given more attention than in most other reviews. Data on mineral matter and trace elements are reviewed, and 116 references to them are included.
11. Breger, Irving A., "Geochemistry of Coal," *Econ. Geol.* (1958) **53**, 823–841.
    This introduction to the geochemistry of coal includes discussions of the origin and the physical and chemical structure of coal in addition to a discussion of the composition of coal.

Literature Cited

1. Rao, C. Prasada, Gluskoter, H. J., "Occurrence and Distribution of Minerals in Illinois Coals," *Ill. State Geol. Surv. Circ.* (1973) **476**, 56 pp.
2. O'Gorman, J. V., Walker, P. L., Jr., "Mineral Matter and Trace Elements in U.S. Coals," Office of Coal Research, U.S. Department of the Interior, Research and Development Report No. **61**, Interim Report No. 2 (1972) 184 pp.
3. Crossley, H. E., "A Contribution to the Development of Power Stations," *J. Inst. Fuel* (1963) **36**, 228–239.
4. Gluskoter, H. J., Simon, J. A., "Sulfur in Illinois Coals," *Ill. State Geol. Surv. Circ.* (1968) **432**, 28 pp.
5. Pringle, W. J. S., Bradburn, E., "The Mineral Matter in Coal. II—The Composition of the Carbonate Minerals," *Fuel* (1958) **37**, (2), 166–180.
6. Dixon, K., Skipsey, E., Watts, J. T., "The Distribution and Composition of Inorganic Matter in British Coals. Part 3: The Composition of Carbonate Minerals in the Coal Seams of the East Midlands Coalfields," *J. Inst. Fuel* (1970) **43**, (354), 229–233.
7. Kemežys, M., Taylor, G. H., "Occurrence and Distribution of Minerals in Some Australian Coals," *J. Inst. Fuel* (1964) **37**, (284), 389–397.
8. "Annual Book of ASTM Standards," Part 19, pp. 438–439, ASTM Std. D1374-73, American Society for Testing and Material, Philadelphia, 1973.
9. Parr, S. W., "The Classification of Coal," *Univ. Ill. Eng. Exp. Sta. Bull.* (1928) **180**, 62 pp.
10. King, J. G., Maries, M. B., Crossley, H. E., "Formulae for the Calculation of Coal Analyses to a Basis of Coal Substance Free of Mineral Matter," *J. Soc. Chem. Ind. London* (1936) **57**, 277–281.
11. Brown, R. L., Caldwell, R. L., Fereday, F., "Mineral Constituents of Coal," *Fuel* (1952) **31**, (3), 261–273.
12. Millot, J. O., "The Mineral Matter in Coal. I—The Water of Constitution of Silicate Constituents," *Fuel* (1958) **37**, (1), 71–85.
13. Given, P. H., "Problems of Coal Analysis," Pennsylvania State Univ. Rept. **SROCR-9**, submitted to the U.S. Office of Coal Research under Contract No. **14-01-0001-390**, 40 pp., 1969.
14. Gluskoter, H. J., "Electronic Low-temperature Ashing of Bituminous Coal," *Fuel* (1965) **44**, (4), 285–291.
15. Gluskoter, H. J., "Clay Minerals in Illinois Coals," *J. Sediment Petrology* (1967) **37**, (1), 205–214.
16. Estep, P. A., Kovach, J. J., Karr, C., Jr., "Quantitative Infrared Multicomponent Determinations of Minerals Occurring in Coal," *Anal. Chem.* (1968) **40**, (2), 358–363.
17. Wolfe, D. F., "Noncombustible Mineral Matter in the Pawnee Coal Bed, Powder River County, Montana," M.S. Thesis, Montana College of Mineral Science and Technology, 1969.

18. Warne, S. St. J., "Identification and Evaluation of Minerals in Coal by Differential Thermal Analysis," *J. Inst. Fuel* (1965) **38**, (292), 207–217.
19. Warne, S. St. J., "The Detection and Identification of the Silica Minerals Quartz, Chalcedony, Agate, and Opal, by Differential Thermal Analysis," *J. Inst. Fuel* (1970) **43**, (354), 240–242.
20. Dutcher, R. R., White, E. W., Spackman, W., "Elemental Ash Distribution in Coal Components—Use of the Electron Probe," *Proc. 22nd Ironmaking Conf., Iron Steel Div., Metallurgical Soc., Amer. Inst. Mining Eng.,* New York (1964) 463–483.
21. Gluskoter, H. J., Ruch, R. R., "Iron Sulfide Minerals in Illinois Coals," *Geol. Soc. Amer. Abstr.* **3**, 582.
22. Gluskoter, H. J., Lindahl, P. C., "Cadmium: Mode of Occurrence in Illinois Coals," *Science* (1973) **181**, (4096), 264–266.
23. Ruch, R. R., Gluskoter, H. J., Shimp, N. F., "Occurrence and Distribution of Potentially Volatile Trace Elements in Coal," *Ill. State Geol. Survey Environ. Geol. Note* (1973) **61**, 43 p.
24. Lefelhocz, J. F., Friedel, R. A., Kohman, T. P., "Mössbauer Spectroscopy of Iron in Coal," *Geochim. Cosmochim. Acta* (1967) **31**, (12), 2261–2273.
25. Rafter, T. A., "Sulfur Isotope Measurements on New Zealand, Australian, and Pacific Islands Sediments," *Proc. Nat. Sci. Found. Symp.*, Yale University (April 1962) 42–60.
26. Smith, J. W., Batts, B. D., "The Distribution and Isotopic Composition of Sulfur in Coal," *Geochim. Cosmochim. Acta* (1974) **38**, 121–133.
27. McCartney, J. T., Ergun, S., "An Automated Reflectance Scanning Microscope System for Study of Coal Components. Application to Analysis of Pyrite Distribution," U.S. Bureau of Mines, unpublished data.
28. Liebig, J., Kopp, H., "Jahresbericht über die Fortschritte der Reinen, Pharmaceutischen, und Technischen Chemie," "Physic, Mineralogie, und Geologie für 1847–1848," p. 1120, Giessen, 1849.
29. Swanson, V. E., "Composition of Coal, Southwestern United States," U.S. Geological Survey, Southwest Energy Study, Coal Resources Work Group, Part II, 61 pp., 1972.
30. Kessler, T., Sharkey, A. G., Jr., Friedel, R. A., "Analysis of Trace Elements in Coal by Spark-source Mass Spectrometry," *U.S. Bur. Mines Rept. Invest.* (1973) **7714**, 8 pp.
31. Goldschmidt, V. M., "Rare Elements in Coal Ashes," *Ind. Eng. Chem.* (1935) **27**, 1100–1102.
32. Goldschmidt, V. M., in "Geochemistry" (A. Muir, Ed.), 730 pp., Clarendon, Oxford, 1954.
33. Nicholls, G. D., "The Geochemistry of Coal-bearing Strata," "Coal and Coal-bearing Strata" (D. G. Murchison and T. S. Westoll, Eds.), pp. 267–307, Oliver and Boyd, Edinburgh and London, 1968.
34. Horton, L., Aubrey, K. V., "The Distribution of Minor Elements in Vitrain: Three Vitrains from the Barnsley Seam," *J. Soc. Chem. Ind., London* (1950) **50** (suppl. issue 1), 541–548.
35. Zubovic, P., "Minor Element Content of Coal from Illinois Beds 5 and 6 and Their Correlatives in Indiana and Western Kentucky," U.S. Geological Survey, open file report, 79 pp., 1960.
36. Zubovic, P., Stadnichenko, T., Sheffey, N. B., "The Association of Minor Elements with Organic and Inorganic Phases of Coal," *U.S. Geol. Survey Prof. Paper* (1960) **400-B**, B84–B87.
37. Zubovic, P., Stadnichenko, T., Sheffey, N. B., "Chemical Bases of Minor Element Associations in Coal and Other Carbonaceous Sediments," *U.S. Geol. Surv. Prof. Pap.* (1961) **424-D** (Article 411) D345–D348.

38. Zubovic, P., Stadnichenko, T., Sheffey, N. B., "Distribution of Minor Elements in Coal Beds of the Eastern Interior Region," *U.S. Geol. Surv. Bull.* (1964) **1117-B**, 41 pp.
39. Zubovic, P., "Physiochemical Properties of Certain Minor Elements as Controlling Factors of Their Distribution in Coal," ADVAN. CHEM. SER. (1966) **55**, 221–246.
40. Ruch, R. R., Gluskoter, H. J., Shimp, N. F., "Occurrence and Distribution of Potentially Volatile Trace Elements in Coal," *Ill. State Geol. Surv. Environ. Geol. Note* (1974) **72**, 96.
41. Bohor, B. F., Gluskoter, H. J., "Boron in Illite as a Paleosalinity Indicator of Illinois Coals," *J. Sediment. Petrology* (1973) **43**, (4), 945–956.
42. Ruch, R. R., Gluskoter, H. J., Kennedy, E. J., "Mercury Content of Illinois Coals," *Ill. State Geol. Survey Environ. Geol. Note* (1971) **43**, 15 pp.
43. Watt, J. D., "The Physical and Chemical Behaviour of the Mineral Matter in Coal under the Conditions Met in Combustion Plant, Part I. The Occurrence, Origin, Identity, Distribution and Estimation of the Mineral Species in British Coals," Literature Survey, 121 pp., British Coal Utilization Research Association, Leatherhead, Surrey, England, 1968.

RECEIVED April 11, 1974

Trace Impurities in Coal by Wet Chemical Methods

EUGENE N. POLLOCK

Ledgemont Laboratory, Kennecott Copper Corp., Lexington, Mass. 02173

In determining trace elements in coal by wet chemical methods, conventional atomic absorption spectroscopy (AAS) was used to determine Li, Be, V, Cr, Mn, Co, Ni, Cu, Zn, Ag, Cd, and Pb after dry ashing and acid dissolutions. A graphite furnace accessory was used for the flameless AAS determination of Bi, Se, Sn, Te, Be, Pb, As, Cd, Cr, Sb, and Ge. Mercury can be determined by flameless AAS after oxygen bomb combustion. Arsenic and antimony can be determined as their hydrides by AAS after low temperature ashing. Germanium, tin, bismuth, and tellurium can be determined as their hydrides by AAS after high temperature ashing. Selenium can be determined as its hydride by AAS after a special combustion procedure or after oxygen bomb combustion. Fluorine can be determined by specific ion analysis after oxygen bomb combustion. Boron can be determined colorimetrically.

As trace impurities enter the environment in increasing quantities, the materials that can have an important impact on the environment are coming under careful scrutiny. Coal and other energy sources are of major interest because they contain elements that can have undesirable physiological effects on plant and animal life, such as Hg, Be, Se, As, Cd, Pb, and F.

Increased environmental concern has accelerated research on the analysis of trace elements in fuels in many university and governmental facilities. Because instruments such as mass spectrometers and nuclear reactors for neutron activation analysis are available, much of this research uses sophisticated instrumentation and techniques. However, the wet chemistry laboratory is still the only available source of chemical analysis capability for the service laboratory of a coal mining facility, even a very large coal mining company.

The coal service laboratory, which in the past was concerned with determining the ASTM Procedures for Ultimate and Proximate Analysis, is now responsible for analyzing trace elements in coal. With this sort of facility and technical skill in mind, our own wet chemical laboratory devised relatively routine procedures for determining trace elements in coal that could have an undesirable environmental impact including Hg, Be, Se, As, Cd, Pb, F, Cu, Ni, Zn, Cr, Te, Ge, Mn, Sn, B, Bi, Sb, V, Li, Co, and Ag.

Instrumentation for a Wet Chemical Laboratory

All of the instrumentation and the analytical chemistry skills that are needed for trace element analysis are essentially present in the normal wet chemistry laboratory. Necessary equipment includes the following:

1. Spectrophotometer
2. pH Meter, expanded scale with fluoride electrode
3. Parr oxygen bomb
4. Atomic absorption spectrophotometer (AAS) with accessories
5. Furnace for high temperature ashing (HTA)
6. Specialized glassware (minor in cost)
7. General miscellaneous glassware and appurtenances of the classical wet laboratory.

While our research was concerned with developing wet chemical methods, we confirmed our data with analyses from an available spark source mass spectrometer (SSMS). The SSMS operating parameters are given in Table I. The instrument used was an AEI MS-7 (1, 2) equipped with electrical detection. It was used in the peak switching mode only to provide more precise analyses.

Sample Preparation

Samples for SSMS were prepared from representative portions of the same HTA-prepared samples used in the wet chemical analyses. The HTA procedure is the following.

Weigh 5–6 g < 100 mesh coal into a porcelain crucible and place it in cold vented furnace. Elevate the temperature to 300°C for 0.5 hr, to 550°C for 0.5 hr, and finally to 850°C for 1.0 hr. Remove the crucible from the furnace and stir the ash with a rod. Return the crucible to the furnace at 850°C for 1.0 hr with no venting.

The SSMS samples are reground with a boron carbide mortar and pestle and diluted with two parts of high purity graphite. The samples with graphite are placed in polystyrene vials with two or three ⅛-in.

Table I.   SSMS Operating Parameters

Spark variac	35%
Pulse repetition rate (pps)	100
Pulse length (μ sec)	100
Source slit	0.002 in.
Multiplier slit	0.002 in.
Monitor exposure	0.3 nanocoulombs
Multiplier and amplifier gains	⎡ Variable according ⎤
Electrodes, vibrated	⎢ to sample elements ⎥
	⎣ and concentration ⎦

polystyrene beads and mixed in a spex mill for 20 min. Electrodes are prepared from the powders using the AEI briquetting die and polyethylene slugs.

Table II compares SSMS (G) results and conventional solution atomic absorption data from coals ashed as described above. The data indicates reasonable if not outstanding agreement especially since the coals were ashed separately for the SSMS and the AAS samples.

Table II.   Comparison of AAS and SSMS (G) (ppm)

Cu		Zn		Mn	
SSMS	*AAS*	*SSMS*	*AAS*	*SSMS*	*AAS*
14	17	13	9	33	37
23	16	41	45	200	193
22	18	17	12	90	65
33	29	22	24	101	93
11	13	41	39	62	52
32	29	84	12	124	96
12	24	8	13	11	19
38	61	660	852	143	117
13	14	18	20	22	21
18	18	50	36	73	49

Ni		Cr		V	
SSMS	*AAS*	*SSMS*	*AAS*	*SSMS*	*AAS*
14	23	26	24	25	32
23	30	20	12	21	21
21	31	21	17	35	23
14	21	12	16	43	24
14	22	18	19	24	25
28	28	19	12	24	21
7	15	13	13	15	24
41	81	36	31	20	40
7	22	15	20	61	37
15	18	37	16	27	30

The selection of the correct dissolution step in coal decomposition is vital in determining trace elements. Such elements as copper and nickel can easily be picked up as contaminants from the laboratory environment or reagents. Other elements such as mercury and selenium can be lost in the dissolution step. The dissolution procedure involving the least exposure to contamination without potential loss of volatile components should be used in for each trace element.

Dry ashing is still the simplest prior treatment and should be used where high temperature ashing is feasible. A comparison of wet ashing and HTA was made for nine elements. The results indicate no appreciable loss from volatilization (Table III).

Table III. Comparison of HTA and Wet Ashing[a]

Cu		*Mn*		*Ni*		*Zn*		*Cd*	
HTA	*Wet*	*HTA*	*Wet*	*HTA*	*Wet*	*HTA*	*Wet*	*HTA*	*Wet*
17	16	67	64	19	15	121	110	2.8	2.8
12	15	179	169	4	6	7	9	0.6	0.6
50	50	122	128	84	85	1420	1450	13.1	12.8
13	14	12	15	13	12	16	23	0.6	20.5
9	9	8	7	7	5	17	19	<0.5	0.5

Pb		*Be*		*Li*		*V*	
HTA	*Wet*	*HTA*	*Wet*	*HTA*	*Wet*	*HTA*	*Wet*
13	15	2.2	2.4	6.8	8.0	55	41
8	16	0.4	0.7	9.5	11.2	10	10
105	111	2.3	2.6	17.1	20.8	43	43
13	18	1.6	2.2	11.8	12.8	18	22
8	8	0.3	0.8	3.1	3.6	7	10

[a] AAS data in ppm.

Procedures for Trace Element Analysis

The wet chemical methods described below can be divided into four major groups: elements that can be determined by conventional flame AAS (Li, Be, V, Cr, Mn, Ni, Co, Cu, Zn, Ag, Cd, and Pb), elements that can be determined by flameless AAS (Bi, Zn, Sn, Te, As, Be, Pb, Cd, Cr, Sb, and Ge), elements that can be determined by AAS after the evolution of their volatile hydrides (5, 6) (As, Sb, Bi, Te, Se, Ge, and Sn), and a miscellaneous group (Hg, Se, F, B) which require special methods.

Determination of Li, Be, V, Cr, Mn, Ni, Co, Cu, Zn, Ag, Cd, and Pb. Coal samples for conventional flame AAS can be prepared by either wet ashing or by HTA. The wet ashing procedure uses approximately 5 g of coal treated with a mixture of ammonium persulfate, nitric acid, and sulfuric acid. In the HTA procedure, 5 g of coal were ashed as described in the procedure used for the SSMS (G) analysis.

The ash is placed in a 100 ml Teflon beaker, containing 5 ml of HF (conc.) and 15 ml of HNO_3 (conc.). Dissolve the ash by warming, then evaporate just to dryness. Add water and 1 ml of HNO_3 to dissolve the salts and transfer to a 100 ml volumetric flask. Make to volume with water and mix. Immediately transfer to a plastic bottle and preserve as a stock solution for determining elements which do not require a specialized dissolution technique.

Li, Be, V, Cr, Mn, Ni, Co, Cu, Zn, Ag, Cd, and Pb can be determined by direct flame atomization. If the sample concentration is not in the direct working range, dilution is the preferred method for decreasing the sensitivity of the element. The standard operating procedure using air–acetylene, as recommended by the AAS instrument manufacturer, is adequate for most elements. Beryllium and vanadium are aspirated best with a nitrogen oxide–acetylene flame. The analysis of chromium is optimized with a nitrogen oxide–acetylene flame in the presence of 2% potassium bisulfate. Aqueous standards containing approximately the same amounts of dissolving acids and interference suppressants are used for comparison.

Determination of Bi, Sn, Te, As, Be, Pb, Cd, Cr, Sb, and Ge. For laboratories having a graphite furnace or some similar AAS accessory, flameless AAS provides greater sensitivities than conventional flame procedures for Be, Cr, Pb, or Zn and greater convenience than hydride volatilization procedures (5, 6) for As, Bi, Sn, Te, Sb, or Ge. An electrically heated graphite furnace, the Perkin-Elmer HGA 2000, was used with a Perkin-Elmer AAS model 503. Stock solutions from the HTA (nitric acid–hydrofluoric acid) dissolution were introduced in 50 μl quantities and ignited under conditions indicated in Table IV. Standard solutions for these elements were treated similarly.

Determination of As, Sb, Bi, Te, Ge, and Sn (5, 6). Place coal ash from the HTA procedure in a 100 ml Teflon beaker containing 5 ml of HF (conc.) and 15 ml of HNO_3 (conc.) Warm to dissolve the ash and carefully evaporate to dryness. Add water and 1 ml of H_2SO_4 (conc.) to dissolve the salts and transfer to a 100 ml volumetric flask. Make to volume with water and mix. Immediately transfer to a plastic bottle and preserve as a stock solution for those elements that can be determined by hydride evolution. The hydrides of arsenic, antimony, bismuth, and tellurium are formed by the reaction of nascent hydrogen generated by magnesium metal in a $TiCl_3$ solution. A Perkin-Elmer high sensitivity arsenic-selenium sampling system modified by replacement of the zinc metal port with a simple 3-in. length of rubber tubing and a pinch clamp can be used.

Alternatively, the system in Figure 1 can be used with the following directions for arsenic, antimony, and bismuth.

Table IV.   Parameters for the Graphite Furnace for Coal Samples[a]

| Element | Charring | | Atomization | | | |
	Temp (°C)	Time (sec)	Temp (°C)	Time (sec)	WL nm	Special Condition[b,c]
As	1450	20	2500	10	193.7	D_2, EDL
Be	1200	40	2600	10	235	D_2
Bi	850	30	2300	10	223	D_2
Cd	400	20	2100	10	229	D_2, EDL
Pb	500	10	2100	10	283	
Sb	800	30	2400	10	218	D_2
Se	1250	30	2700	8	196	D_2, EDL, Ni (200 ppm)
Sn	800	10	2500	10	286	
Ge	1000	40	2700	8	265	
Te	350	20	2700	10	213	D_2, EDL

[a] All samples were dried at 125°C for 40 secs.
[b] D_2 = deuterium background corrected.
[c] EDL = electrodeless discharge lamp.

Add an aliquot of arsenic containing up to 0.3 μg, or an aliquot of up to 0.6 μg of antimony or bismuth to a 125 ml erlenmeyer flask containing a Teflon-coated stirring bar. Adjust the volume to 15 ml with water. Add 10 ml of TiCl₃ (1%) in HCl (12N) and let stand for at least 5 min. Connect the flask to the hydride generating system and start the stirrer. With the atomic absorption spectrophotometer on, including the triple slot argon entrained hydrogen burner and a properly zeroed recorder, open the solenoid valve and flush the system for 15 sec. Shut the solenoid valve and drop a 1-in. length of ⅛-in. magnesium rod into the reaction flask through the reagent orifice. Set a timer for 1 min and collect the H_2 and volatile metal hydrides. After 1 min, open the compressed air toggle valve (preset at 5 lbs) to pressurize the pressure chamber. Open the solenoid valve to expel the generated gases into AAS triple slot burner while recording the peak response on the recorder. Close the solenoid valve, shut off the compressed air, and depressurize the pressure chamber through the pressure release valve. A deuterium background accessory was used.

The procedure for tellurium is similar to that used for the previous elements except that there is no time delay in the preliminary reduction step, the collection time is 15 sec, and only 6 ml of titanium trichloride solution is used to optimize the determination.

Germanium and tin can also be determined as their hydrides. Appropriately sized aliquots (< 1.0 ppm tin and < 10 ppm germanium) are placed in the 125 ml erlenmeyer reaction flask and diluted to 20 ml with water and 5 ml of H_2SO_4 (12N). Add a Teflon-coated stirring bar and connect the flask into the hydride generating system. Start the stirring and flush the system for 15 sec with argon. Close the flushing system and add a 0.2–0.3 g pellet of $NaBH_4$ through the reagent orifice. Collect the gases for 10 sec, then pressurize the pressure chamber. At the 15-sec mark, open the exit valve expelling the H_2 and metal hydrides into the AAS triple slot burner. Compare the germanium or tin content with standards.

Time factors and reagent parameters are the same for As, Sb, Bi, Te, Ge, and Sn when using the modified Perkin-Elmer high sensitivity arsenic–selenium sampling system. The sodium borohydride reduction can be used for arsenic, antimony, bismuth, and tellurium as well.

Determination of Mercury, Fluorine, Boron, and Selenium. THE DETERMINATION OF MERCURY. A coal sample is decomposed by igniting a combustion bomb containing a dilute nitric acid solution under 24 atm of oxygen. After combustion, the bomb washings are diluted to a known volume, and mercury is determined by atomic absorption spectrophotometry using a flameless cold vapor technique.

Transfer approximately 1 g of 100 mesh XO coal to a clean combustion crucible and weigh to the nearest 0.1 mg. Transfer 10 ml of 10%

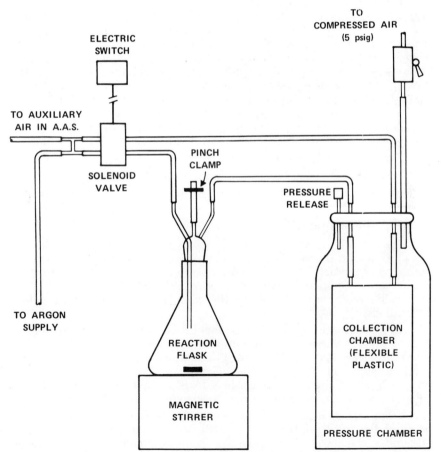

Figure 1.   Hydride generator and sampling system

HNO_3 to the bomb, place the crucible in the electrode support of the bomb, and attach the fuse wire. Assemble the bomb and add oxygen to a pressure of 24 atm (gage). Place the bomb in the calorimeter (a cold water bath in a large stainless steel beaker is also satisfactory) and ignite the sample using appropriate safety precautions ordinarily employed in bomb calorimetry work.

After combustion, the bomb should be left undisturbed for 10 min to allow temperature equilibration and the absorption of soluble vapors. Release the pressure slowly and transfer the contents of the bomb (and crucible) to the mercury reduction vessel by washing with 10% HNO_3. Rinse the bomb, electrodes, and crucible thoroughly with several small washings of 10% HNO_3. Dilute the contents of the reduction vessel with 10% HNO_3 to a total volume of 50 ml. Proceed with the determination as described under in the stanardization section below. Determine the amount of mercury in μg and divide by the sample weight in grams to obtain the mercury value in ppm.

This procedure is used for standardization. Add an aliquot of standard mercury solution containing 0.1 μg of mercury to the mercury reduction flask. Add $KMnO_4$ (3%) drop-wise until the pink color persists. Adjust the volume to 50 ml, then add in order: 5 ml of HNO_3 (1:2), H_2SO_4 (1:1), and hydroxylamine hydrochloride. When the pink color fades, add 5 ml of the $SnCl_2$ (10%) and immediately connect into the system. Start the pump which circulates the mercury in the vapor phase through the optical cell in the atomic absorption spectrophotometer with the mercury lamp optimized at 253.7 nm and normal operating conditions as established by the AA instrument manufacturer. Samples are run by taking all or an aliquot from the oxygen bomb combustion stock solution.

THE DETERMINATION OF FLUORINE. A coal sample is decomposed by ignition in a combustion bomb containing Na_3CO_3 solution under 24 atm oxygen. After combustion, the bomb washings are diluted to a known volume and an aliquot is taken to determine fluorine by the standard Orion specific ion procedure.

Transfer approximately 1 g of coal to a combustible crucible and weigh to nearest 0.1 mg. Transfer 5 ml of Na_2CO_3 solution (5%) to the bomb, place a crucible in the electrode support, and attach the fuse wire. Assemble the bomb and oxygen to a pressure of 24 atm. Place the bomb in a cold water bath and ignite the sample using appropriate safety precautions.

After combustion, the bomb should be left undisturbed for 10 min to allow temperature equilibration and absorption of the soluble vapors. Release the pressure slowly and transfer contents of the bomb (and crucible) into a 25 ml volumetric flask. Make several small washings with water and add the rinsings to the volumetric flask. Make to volume with water and reserve the stock solution in a plastic bottle.

Using an expanded scale pH meter, such as the Orion 801, pipet 10 ml of the stock solution into a small beaker and add 10 ml of Tisab (Orion No. 94-09-09). Determine the electrode potential using a fluoride electrode Orion 94-09. Comparison is made by bracketing with fluoride standards prepared similarly.

THE DETERMINATION OF BORON. After dry ashing in the manner used in the lithium, beryllium, etc., procedure, the ash is fused with sodium carbonate, leached in water, and acidified with sulfuric acid. The colorimetric carminic acid method is then used to determine the boron.

Weigh approximately 1 g of coal into a platinum crucible and carefully ignite in a vented oven. Gradually increase the temperature to 850°C and maintain for 1 hr. Remove from the oven and add 2 g of Na_2CO_3 and fuse for 10 min. Leach in 25 ml of warm water in Teflon beaker. When dissolution is complete, carefully add 10 ml of H_2SO_4. Transfer to a 50 ml volumetric flask. Make to volume with water and reserve in a plastic bottle.

Place a 5 ml aliquot or less in a 50 ml volumetric flask and make to 5 ml volume with H_2SO_4 ($3.6M$) if less than a 5 ml aliquot is used. The aliquot should contain 0–100 μg of boron. Add 20 ml of chilled H_2SO_4 ($18M$) and swirl. Then, by pipet, add 20 ml of carminic acid (0.92 g in H_2SO_4 ($18M$)). Make to volume with H_2SO_4 ($18M$) and determine the absorbance in 1 cm cells in a spectrophotometer at 605 nm with a reagent blank in the reference cell. Compare with a standard curve containing 0–100 μg of boron.

THE DETERMINATION OF SELENIUM. The most difficult trace element to determine in coal by wet chemical methods is selenium. Two alternative dissolution techniques can be used—H. L. Rook's combustion method (7) and the oxygen bomb combustion method (4). Also, two alternative analytical methods can be used—the hydride evolution method (5) and the graphite furnace method.

To determine selenium in coal by ignition in a Parr oxygen bomb containing water, transfer approximately 1.0 g of 100 mesh coal to a clean combustion crucible and weigh the coal to the nearest 0.1 mg. Transfer 10 ml water to the bomb, place the crucible in the electrode support of the bomb, and attach the fuse wire. Assemble the bomb and add oxygen to a pressure of 24 atm (gage). Place the bomb in a cold water bath and ignite the sample using appropriate safety precautions ordinarily employed in oxygen bomb combustion work. After combustion, the bomb should be left undisturbed for 10 min to allow for temperature equilibration and absorption of soluble vapors. Release the pressure slowly and transfer the contents of the bomb and crucible into a 150 ml beaker while filtering out the solids.

To provide a stock solution for flameless atomic absorption, add 25 ml of HCl (conc.) and make the volume to approximately 50 ml with water. Add 5 ml 0.125% arsenic solution (5.0 mg) and heat to boiling. Add 8 ml of hypophorous acid (50%) and boil for 5 min. Cool and filter through two 5μ Teflon millipore filters. Wash with HCl (1:1) and finally with water. Place millipore funnel and Teflon filters in a 50 ml beaker. Add two 1 ml portions of HNO_3 down the sides of the millipore funnel. Then heat to complete the dissolution of selenium and arsenic. Rinse off the funnel and millipore filters. Transfer to a 25 ml volumetric flask and add 10 mg. of nickel as $Ni(NO_3)_2$. Make to volume with water.

This provides a stock solution for flameless atomic absorption using 50 μl aliquots.

The system used was a Perkin-Elmer model 503 atomic absorption spectrophotometer and a Perkin-Elmer graphite furnace model HGA 2000. Evaporation temperature was 125°C, charring temperature was 1250°C, and atomization temperature was 2700°C. Comparison was made to 25, 50, and 100 ppb selenium standard solutions in 8% HNO_3 and 400 ppm of $Ni(NO_3)_2$.

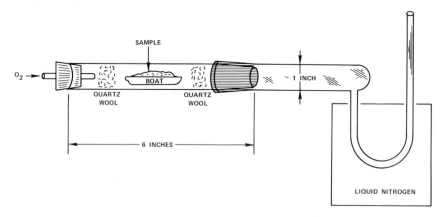

Figure 2. Selenium combustion apparatus

Alternatively, the bomb combustion dissolution procedure can be combined with the hydride evolution method previously described for arsenic, tin, and bismuth. After completion of the bomb combustion and absorption of gases into the 10 ml of water, the solids are filtered out and the filtrate collected into a 50 ml volumetric flask containg 20 ml of hydrochloric acid. This is a suitable stock solution for the hydride evolution method.

Selenium can also be separated from the bulk of the coal samples by the combustion technique described by H. L. Rook (7), which was originally used in a neutron activation analysis. The equipment was

Table V.   Comparison of Graphite

	Be		Pb		As	
	Furnace	Std AAS	Furnace	Std AAS	Furnace	Std AAS
DRB A	1.52	1.5	9.8	14	14.6	—
B	1.60	1.6	14.7	17	24.8	—
C	1.50	1.6	7.4	9	32.1	—
D	2.71	2.9	13.4	11	5.5	—
NBS Coal	1.24	1.4[a]	30.8	30[a]	5.7	5.9[a]
Fly ash	9.56	12[a]	55	70[a]	61.2	61

[a] NBS, Neutron activation data.

modified to use a quartz combustion tube 5½ in. long connected by a ground glass joint to a second section 3½ in. long. The section ends with a U-tube which can be immersed in a dewar flask containing liquid nitrogen gas or some other suitable coolant (Figure 2).

For analyses, an 0.5 g sample is weighed into a porcelain boat and inserted into the combustion tube with an oxygen flow of ~30 cc/min. With a cold trap in place, the sample is ignited by heating the combustion tube with a meker burner. The coal sample is allowed to burn freely, and then the temperature is raised to the burner maximum for 5 min. The combustion tube is cooled for 5 min and separated from the condenser section. The condenser is removed from the cold trap and allowed to warm to ambient temperature. Add 10 ml of water to the condenser and flush into a 50 ml volumetric flask. Make to volume with water and mix. Take an aliquot of 15 ml or less containing up to 0.3 μg of selenium and proceed as in the hydride method for arsenic, tin, and bismuth by AAS, as previously described.

The graphite furnace method can also be used as described in the bomb combustion procedure. Table V compares graphite furnace values with conventional flame AAS or NBS neutron activation analyses.

Conclusions

As a result of this investigation of wet chemical procedures, I believe that a coal service laboratory, with a modest increase in instrumentation, can essentially determine all the significant trace elements with the necessary precision and sensitivity.

Acknowledgments

I wish to acknowledge the aid of James R. Jones of Peabody Coal Co. in writing the introductory comments to this paper. I also wish to acknowledge the use of the spark source mass spectrometry data obtained from Richard J. Guidoboni of Kennecott Copper Corp., Ledgemont Laboratory.

Furnace with Other Methods (ppm)

Cd		Cr		Sb	Ge
Furnace	*Std AAS*	*Furnace*	*Std AAS*	*Furnace*	*Furnace*
0.21	1.2	19.1	19	0.14	24.9
0.17	0.8	19.7	19	0.73	26.3
0.14	0.8	15.9	11	1.03	21.1
0.14	0.5	15.2	11	0.42	16.1
0.25	0.24[a]	19.4	20[a]	3.62	14.9
1.45	1.45[a]	181	132	17.4	131.0

Literature Cited

1. Ahearn, A. J., Ed., "Trace Analysis by Mass Spectrometry," Academic, New York, 1972.
2. Evans, Jr., C. A., Guidoboni, R. J., Leipziger, F. D., *Appl. Spectrosc.* (1970) **24**, 85.
3. Kometani, T. Y., *et. al.*, *Environ. Sci. Technol.* (1972) **6**, 617.
4. *Book ASTM Stand.*, ASTM **D808** Part 17.
5. Pollock, E. N., West, S. J., *At. Absorption Newslett.* (September–October, 1972) **11**, (5).
6. *Ibid.*, (January–February, 1973) **12**, (1).
7. Rook, H. L., *Anal. Chem.* (June 1972) **44**, (7), 1276.
8. *Orion Bull.* (1968) (5).
9. Pollock, E. N., Zopatti, L. P., *Talanta* (1963) **10**, 118.

RECEIVED January 21, 1974

Trace Elements in Coal
by Optical Emission Spectroscopy

G. B. DREHER and J. A. SCHLEICHER

Illinois State Geological Survey, Urbana, Ill. 61801

Two quantitative emission spectrochemical methods—direct-reading and photographic—have been developed to analyze 16 trace elements in high-temperature coal ash. Problems such as possible loss of vanadium and molybdenum by volatilization and an apparent increase in boron concentration with increasing ashing temperature were encountered in ashing coal samples. Because natural and/or analyzed standards were unavailable, synthetic standards were prepared for the direct-reading method. U.S.G.S. standard granite G-2 and iron oxide were combined with various amounts of a mixture of silica, alumina, and 1000 ppm of each of the trace elements of interest. For the photographic method, similar mixtures were prepared without the G-2 base. For each element, usable concentration limits, concentration ranges on the coal basis calculated from the coal ash samples, and the resulting standard deviations were determined.

Environmental concern in recent years has caused increasing awareness of the small quantities of substances in our air, water, and food which can affect living things. It has become necessary to know the sources of these potentially toxic substances. Coal, which is burned in large quantities annually in the U.S. to produce energy, is one such source because it contains trace concentrations of many possibly toxic elements that may be released to the environment in considerable quantities. To assess this possibility, concentrations of these trace constituents must be determined in the whole coal before combustion. Reliable and accurate analysis methods are required to produce the data that ultimately may be used by those charged with protecting our environment to set standards and by coal users to see that those standards are met.

Methods have been developed at the Illinois State Geological Survey to determine the concentrations of 23 potentially volatile trace elements in coal. Some of the concentrations were checked by two or more analytical techniques. Analyses have been made by five different techniques: neutron activation analysis, atomic absorption spectrometry, x-ray fluorescence spectrometry, optical emission spectroscopy (direct-reading and photographic), and ion specific electrode. The method used determines whether whole coal, low-temperature ash, or high-temperature ash is used as the analytical sample. Since comparable results for certain elements were produced, there is a high degree of confidence in the methods developed. Of the two optical emission spectroscopic methods described, a direct-reading method was used to analyze Sn, B, Pb, Cu, Co, Ni, Be, Cr, V, Mo, and Ge, and a photographic technique was used for B, Mn, Cr, Pb, Be, V, Ag, Cu, Zn, Zr, Co, and Ni. Sixteen of these elements were determined in the high-temperature ash of the coals, and eight of them were cross-checked by both methods. The silver content in the ash samples analyzed was always very near the detection limit, and although bismuth was sought, it was never detected. Conventional d.c. arc excitation is used by both methods.

Quantitative spectrographic analyses for trace elements in high-temperature ash samples of coal have been reported by Abernethy *et al.* (*1*), Zubovic *et al.* (*2, 3, 4, 5*), Rao (*6*), and Hunter and Headlee (*7*). We felt that it was desirable to develop additional analysis methods, especially for the direct-reading spectrometric technique, in which small changes of some matrix constituents might cause relatively large variations in results because of the increased sensitivity of the detection system. A

Table I.   Concentrations in Standard Materials of Synthetic Standard for

Element (ppm)	Weight of synthetic standard (mg) added to 0.5 g G-2 + 379 mg Al_2O_3			
	0.00	1.00	4.04	9.19
B	1.86	2.84	5.78	10.7
Pb	26.7	28.5	33.9	43.0
Cu	9.95	11.8	17.3	26.6
Co	4.55	6.40	12.0	21.3
Ni	5.95	7.79	13.4	22.6
Be	2.23	4.08	9.67	19.0
Cr	8.37	10.2	15.8	25.0
V	34.4	36.2	41.6	50.6
Sn	0.93	2.78	8.37	17.7
Mo	1.12	2.97	8.57	17.9
Ge	0.66	2.51	8.11	17.5

method for silicate analysis (8) was adapted for the coal ash analyses by the photographic technique.

Experimental

Preparation of High-Temperature Ash. Two grams of finely ground coal (−60 mesh or smaller) are separated from the larger stock by riffling, weighed into a tared, used silica crucible, and dried in an oven at 110°C for approximately 3 hrs. (Gorsuch (9) and Ruch *et al.* (10) showed that the walls of new silica crucibles retain some trace elements to a greater extent during dry ashing than crucibles that have been used for several ashings.) The samples are cooled and weighed for moisture determination. The weighed, moisture-free samples are transferred to a cold muffle furnace, heated to 500°C in 1 hr, and ashed at that temperature until no carbonaceous matter appears. During ashing, the samples are mixed with a platinum wire approximately every hour. The samples are cooled and weighed, ground in a mullite mortar, then dried at 110°C, and stored in a desiccator over $Mg(ClO_4)_2$.

Preparation of Direct-Reading Spectrometric Standards. No certified standards are yet available for trace elements in coal, although the National Bureau of Standards is conducting an interlaboratory evaluation of a composite coal sample for use as a Standard Reference Material (11). It was necessary, therefore, to prepare synthetic standards.

For the spectrometric method, U.S. Geological Survey standard granite G-2 is used as a base for the standard. To 0.5 g portions of G-2 are added 379 mg of Specpure Al_2O_3 (Johnson Matthey Chemicals Ltd., through Jarrell–Ash Division, Fisher Scientific Co., 590 Lincoln St., Waltham, Mass. 02154) to adjust the $SiO_2:Al_2O_3$ ratio to 3:1 and various amounts of a synthetic standard containing 1000 ppm (w/w) of each of

G-2 Base in 3:1 Silica–Alumina Matrix + 1000 ppm
Spectrometric Method

Weight of synthetic standard (mg) added to 0.5 g G-2 + 379 mg Al_2O_3				*Weight of* $Na_2B_4O_7$ · $10H_2O$ (mg)
19.79	*54.45*	*133.22*	*535.89*	*6.32*
20.5	50.2	106	264	1320
61.2	116	220	512	
45.1	101	206	504	
39.9	96.0	202	501	
41.2	97.3	203	502	
37.6	94.0	200	500	
43.6	99.5	205	503	
68.7	123	226	516	
36.4	92.8	199	500	
36.6	92.9	199	500	
36.1	92.5	199	499	

the 11 elements being determined. The mixtures are ground in a mullite mortar and then mixed in plastic vials 1 in. in diameter $\times$ 3 in. deep that contain two plastic balls $1/4$ in. in diameter for 1 hr in a mixer mill. Eight standards are prepared in this manner. A ninth standard is prepared similarly by using 6.32 mg $Na_2B_4O_7 \cdot 10H_2O$ in addition to the 0.5 g of G-2 and 379 mg Al_2O_3 to obtain a boron concentration of 1320 ppm. A summary of the amounts of synthetic standard taken and the resulting trace element concentrations are given in Table I.

An array of standards is prepared so that one series of the nine previously described standards contains a total iron oxide concentration (including that attributable to the G-2) of 10%, a second series of standards contains a total iron oxide concentration of 15%, and so on up to 30% iron oxide, with a total of 45 mixtures.

Final standards are prepared by adding 40 mg of the corresponding mixture to 150 mg of SP-2X graphite powder (Spex Industries, Inc., Box 798, Metuchen, N.J. 08840) and 10 mg of Specpure $Ba(NO_3)_2$ and mixing them well in plastic vials, $1/2$ in. in diameter $\times$ 1 in. deep, containing two plastic balls, $1/8$ in. in diameter. Mixtures of high temperature ash samples are prepared in an analogous manner. Forty mg of finely ground high-temperature ash is mixed with 150 mg of SP-2X graphite powder and 10 Specpure $Ba(NO_3)_2$.

Fifteen mg charges of standard or sample are weighed and loaded into thin-walled crater electrodes $1/8$ in. in diameter, National type L-3979 (Spex Industries, Inc.). Arcing conditions are shown in Table II.

Preparation of Spectrographic Standards. For the spectrographic determinations, high-temperature ash samples were again used; however, analytical procedures were different out of necessity. The microphotometer used is capable of a greater resolution than the direct-reading spectrometer because of its inherent magnification ($17\times$), although the reciprocal linear dispersions of the spectrograph and spectrometer are very nearly the same (5 A/mm). It was found, therefore, that the background effects were less significant in the photographic method, although it is more time consuming.

Accordingly, a set of synthetic standards was prepared by mixing commercial SiO_2 base standards containing 1000 ppm of 49 elements (Spex Industries, Inc. #1006) with corresponding 1000 ppm Al_2O_3 base standards (Spex Industries, Inc. #1007) in the ratio of 3:1 (SiO_2:Al_2O_3). This mixture was then diluted with a 3:1 mixture of Specpure SiO_2 and Al_2O_3 to concentrations of 1000, 500, 250, 100, 50, and 10 ppm. The 10 ppm standard was later found to be below the limit of detectability under the exposure conditions used. To nine parts of each of these mixtures was added one part Fe_2O_3, producing final trace element concentrations of 900, 450, 225, 90, 45, and 9 ppm. These standards were, in turn, mixed in a 1:1 ratio with SP-2X graphite powder containing 0.111% indium as internal standard. However, the internal standard was discontinued because it decreased rather than increased precision, probably owing to the refractory nature of the matrix.

Table II.   Standard Operating Conditions

	Spectrometric	*Spectrographic*
Instrument	Jarrell-Ash Model 750 Atomcounter	Jarrell-Ash 3.4 m Ebert with 3-lens collimating system
Arc current	15 amp (short circuit)	11 amp (true)
Arc voltage		250 VDC
Electrode gap	6 mm	4 mm
Exposure time	65 sec	completion + 10 sec (usually about 150 sec)
Electrodes	sample electrode, National L-3979 counter electrode, National L-4036 (ASTM C-1)	sample electrode, National L-3903 (ASTM C-13) counter electrode, National L-4037(ASTM C-1 with radial tip)
Electrode atmosphere	80% argon, 20% oxygen at 10 SCFH	80% argon, 20% oxygen at 14 SCFH
Sample charge	15 mg	20 mg
Attenuation[a]		neutral density filter, 6% T
Step sector		1:1.585, 8 steps
Entrance slit width	10 μm	25 μm
Exit slit width	50 μm	
Development		3.0 min in Eastman D-19, 30 sec in 2.5% acetic acid, 4.0 min in Eastman fixer

[a] Because the concentrations of most trace elements in the ash were too high to be recorded on a total energy ignition, it was necessary to insert a neutral density filter in the light path after the second lens of the three lens collimating system.

The samples (high-temperature ash) and standards, both diluted with an equal weight of SP–2X graphite powder, were ignited spectrographically under conditions shown in Table II.

Results and Discussion

Two types of coal ash samples have been prepared routinely for analysis at the Illinois Geological Survey. Low-temperature ash samples (*12*), in which the bulk of the mineral matter remains unchanged, are prepared by reaction of the coal with activated oxygen in a radiofrequency field. The effective temperature produced by this device is approximately 150°C. Such samples were unsatisfactory for emission spectroscopic analysis. It is postulated that the presence of largely unaltered mineral matter, such as carbonates, sulfides, and hemihydrated sulfates (*12*), caused the observed nonreproducibility of results. High-temperature ash samples, prepared in a muffle furnace, consisted mainly of oxides of the elements present in the original mineral matter and were used throughout this investigation. Results obtained from high-temperature ash, calculated to the whole coal basis, compare well with corresponding results from low-temperature ash and whole coal where cross-checks are available. (*See* Ref. *10.*)

Variation of Trace Element Concentration with Ashing Conditions. To determine whether or not the concentrations of any of the elements of interest were altered with increasing ashing temperature, samples were ashed in uncovered, used porcelain crucibles at 300, 400, 500, 600, and 700°C until the carbonaceous matter was no longer apparent. Two coals were studied, and trace element analyses were carried out on each of the 10 resulting ash samples. The results are shown in Table III. None of the data exhibit losses or gains of trace element concentration in samples ashed between 300 and 700°C, other than a gain in boron concentration and a possible loss of lead. A loss of lead with increasing temperature is expected. For both coals, the boron concentrations increased with ashing temperature from 300 to 600°C, then dropped slightly between 600 and 700°C. Williams and Vlamis (*13*) noted a similar effect when they ashed plant material with calcium hydroxide in muffle furnaces and also when they heated calcium hydroxide in the same furnaces in the absence of plant material. The increase in boron concentration was greater in uncovered crucibles or dishes than in covered. It was thought that boron vapor from the furnace walls condensed in increasing amounts in the alkaline material as temperatures increased for a given heating time or as heating times increased at a given temperature.

Another pair of coals was chosen to determine whether or not a similar effect took place when coal was ashed without alkaline material.

Table III.   Concentrations (ppm) of Trace Elements Samples Prepared at

Sample Number and Temperature

	C-16317				
Element	*300°C*	*400°C*	*500°C*	*600°C*	*700°C*
B	62	90	104	119	106
Pb	77	61	64	65	64
Cu	21	20	21	21	22
Co	7.9	—	8.5	8.9	8.4
Ni	39	39	36	39	35
Be	2.5	2.5	2.6	2.6	2.6
Cr	21	19	19	20	20
V	42	45	36	40	35
Sn	4.4	<4.2	<4.2	<4.2	<4.2
Mo	12	13	9.3	10	7.6
Ge	16	16	20	17	21

The coals were ashed in a Hoskins electric furnace, Type FD204A, in covered and uncovered platinum crucibles at 300, 400, 500, 600, and 700°C for 20 hrs each at a heating rate of 500°C/hr. The results, listed in Table IV, indicated that the boron concentrations again increased when the coal was ashed in uncovered crucibles. However, in covered crucibles, the boron concentrations remained relatively constant until an ashing temperature of 700°C was reached. Boron concentrations in the samples ashed in the covered crucibles are higher, in most instances than those in the corresponding uncovered samples. This suggests that boron was lost from the samples in the uncovered crucibles more easily at low temperatures than at high temperatures. This may have been caused by competing reactions: a loss mechanism, volatilization and/or retention on or in the crucible walls and a reaction recombining freed boron with the bulk sample. The recombination reaction could become more efficient than the loss mechanism at higher temperatures in the uncovered crucibles, and boron loss in the covered crucibles may be prevented by the covers, which provide sufficient residence time for the recombination reaction to occur. The increase in boron concentration in the covered 700°C ash samples is unexplained.

The data for copper indicates that contamination may have come from the platinum crucibles since similar concentration trends were noted for both covered and uncovered crucibles. Under the chosen ashing conditions, vanadium and molybdenum were apparently consistently volatilized.

These increasing and decreasing concentration trends do not appear to be a function of the spectrometric method. Continuous time *vs.* relative intensity curves were prepared for the iron internal standard

in Moisture-Free Coal From High-Temperature Ash
Various Temperatures

Sample Number and Temperature

C-16030				
300°C	*400°C*	*500°C*	*600°C*	*700°C*
23	31	34	46	40
43	49	46	44	43
17	18	17	18	17
15	19	17	17	19
44	47	47	47	49
2.4	2.5	2.4	2.5	2.4
16	17	17	18	18
37	32	33	30	33
<2.0	<2.0	<2.0	<1.9	<1.9
26	19	19	16	18
9.3	7.5	8.1	8.8	7.3

Table IV. Ashing Temperature Study

	Uncovered				
Element	300°C	400°C	500°C	600°C	700°C
		Sample C-17601			
Sn	<1.9	4.2	2.1	<1.6	<1.6
B	30	32	41	45	41
Cu	16	14	15	17	25
Co	15	15	13	14	13
Ni	39	38	36	37	35
Be	2.6	2.5	2.6	2.5	2.3
Cr	19	17	15	17	17
V	32	24	19	18	18
Mo	11	7.8	5.4	5.5	4.7
Ge	5.4	5.4	5.7	5.3	5.6
		Sample C–17215			
Sn	7.4	7.0	6.8	7.4	8.0
B	48	53	64	64	70
Cu	16	17	18	29	33
Co	5.2	4.9	4.8	4.5	4.4
Ni	12	12	13	11	11
Be	5.4	5.5	5.4	5.3	5.6
Cr	23	22	22	20	20
V	44	36	33	28	28
Mo	7.2	6.2	5.3	4.2	4.0
Ge	14	16	14	16	18

[a] Same muffle furnace is used, concentrations are in ppm, on a moisture free coal basis, and ashing time is 20 hrs, throughout.

line and the boron, vanadium, and molybdenum lines for coal C-17601 ashed at 300 and 700°C, covered and uncovered. The resulting curves for each spectral line were similar regardless of the sample used, and in each case total line emission was attained within the 65 sec exposure time normally used.

Another study of the variation of trace element concentrations with ashing time at 500°C in covered and uncovered platinum crucibles indicated that ashing time had no effect (Table V). As previously noted, the loss of boron from the uncovered crucibles stabilized at a relatively constant concentration in less than 5 hrs. Molybdenum and vanadium, which show losses with increasing temperatures, show no apparent ashing time dependence.

Choice of an Internal Standard. One of the difficulties in the spectrometric trace analysis of coal ash samples, in addition to choosing a suitable comparison standard matrix, is choosing an internal standard. The first choice in both analytical methods was indium, which was used as a constant internal standard added to the graphite powder diluent–buffer. The results obtained had poor reproducibility, as previously

With Platinum Crucibles and Covers [a]

		Covered		
300°C	*400°C*	*500°C*	*600°C*	*700°C*
		Sample C–17601		
<2.4	2.9	3.6	2.5	2.6
39	38	42	43	52
15	15	15	18	28
14	14	14	13	13
39	37	37	36	35
2.8	2.5	2.5	2.5	2.6
15	16	17	17	18
31	24	21	18	17
11	8.4	6.2	5.6	4.0
3.4	5.9	5.5	5.6	5.5
		Sample C–17215		
8.4	7.0	6.7	5.9	7.3
78	65	77	72	93
22	17	17	20	34
6.4	5.0	5.3	4.7	4.6
12	12	13	11	10
6.6	5.5	6.0	5.3	5.6
23	21	21	21	22
48	37	34	27	26
9.8	6.4	6.4	4.7	3.7
18	14	17	15	16

mentioned, and in both analytical methods unstandardized data yielded better precision than indium-standardized data. In the direct-reading procedure, iron was chosen as a variable internal standard. Scott *et al.* (*14*), Mitchell (*15*), and Davidson and Mitchell (*16*), described procedures for trace elements analysis in soils in which the naturally occurring iron oxide is used as a variable internal standard. Their investigations resulted in families of working curves for cobalt and chromium in which the ratio of analysis line to standard line intensity for a given concentration of cobalt or chromium decreased with increasing iron oxide concentration. Under the conditions in column two of Table II, it was found that coal ash analyses showed a similar dependence of analytical line to standard line ratio on iron oxide concentration. However, there was no regular variation of relative intensity of an analytical line alone with increasing iron oxide concentration except in cases of spectral interference, *e.g.*, lead, nickel, and molybdenum at low concentration levels.

Before any standards were run, the response of the spectrometer to various concentrations of iron oxide was determined. The average relative intensities obtained for at least four exposures of standards containing

Table V. Ashing Time *vs.* Trace Element Concentrations for Coal C–17601 (ppm)[a]

Element	Uncovered (hrs)			Covered (hrs)		
	5	10.5	20.75	5	10.5	20.75
Sn	1.8	2.1	<1.6	<1.7	<1.7	<1.7
B	39	38	35	47	42	44
Cu	16	18	16	17	16	17
Co	13	14	14	14	14	15
Ni	36	38	37	37	37	38
Be	2.4	2.5	2.8	2.4	2.5	2.8
Cr	17	19	19	17	17	19
V	20	19	21	22	21	19
Mo	6.2	5.9	6.0	7.5	6.5	6.4
Ge	5.0	5.6	5.5	6.0	6.3	6.0

[a] 500°C HTA, covered and uncovered Pt crucibles, same muffle furnace.

10–30% iron oxide in 5% increments were plotted against iron oxide concentration on log-log graph paper and produced a straight line. This response curve was checked by running several high-temperature ash samples for which the iron oxide concentrations were known. The average relative iron line intensities predicted by the response curve were obtained within experimental error ($\pm 10\%$). The response curve was then used to predict relative intensities for a given iron oxide concentration in each high-temperature ash that contained between 10 and 30% iron oxide, and the appropriate instrumental adjustment was made to standardize the exposure.

The total iron concentrations in the whole coal samples were determined by x-ray fluorescence spectrometry, and the concentrations of iron oxide in the corresponding ash samples were calculated.

The addition of 5% (w/w) of Specpure barium nitrate to the sample–graphite mixture improved the precision of the spectrometric analysis. Use of a small diameter, thin-walled crater electrode further improved the precision by reducing the amount of space covered by arc wander.

Statistical Results from the Analysis Methods. Continuous time *vs.* relative intensity curves were made spectrometrically for each of 11 elements in seven different coal ash samples. The results showed that peak intensities for all the elements in each sample were generally reached between 50 and 60 sec after initiation of the arc. This behavior helps to explain why using iron as the variable internal standard was successful for the normally wide range of volatilities represented. The large dilution factor involved or a possible carrier distillation effect of barium nitrate might explain the almost complete absence of fractional volatilization.

Table VI shows the analytical wavelengths, the concentration ranges encountered in the coals, average relative standard deviations, the relative standard deviation ranges, and the detection limits in the sample charge determined for the spectrometric method. In most cases, these statistics are based on four determinations for each of 98 samples. The generally poor precision for the tin data is probably attributable to the fact that tin concentrations were only slightly above or below the detection limit (~6 ppm in the sample charge).

Table VII shows the analytical wavelengths chosen, the concentration ranges calculated to the whole coal, average relative standard deviations, and detection limits in ash determined by the spectrographic method.

In the photographic procedure, the lack of a suitable internal standard for exposure correction, the attempt to record and determine all elements on one generalized exposure, and the very high concentration of the trace elements in the ash (for some samples as much as 33 times the amount reported in the coal) caused a poor relative standard deviation. However, of the 13 elements determined, only Co, Ni, Cr, and V were less precise than ±20%, a level which we feel is suitable for a photographic method.

Gluskoter and Lindahl (*17*), Clark and Swaine (*18*), and Abernethy and Gibson (*19*) have found discrete mineral particles (such as sphalerite, beryl, or zircon) in whole coal samples. The presence of these mineral particles in a coal sample prior to high-temperature ashing could account for a higher variation in the analyses by either method of certain elements in the ashes because of inhomogeneous distribution of the resulting oxides, even after the coal ash has been ground. An analytical

Table VI. Statistical Results of the Spectrometric Method

| | | Coal | | | Ash |
| | | Con-centration Range (ppm) | Average Relative Standard Deviation (%) | Relative Standard Deviation Range (%) | Detection Limit in Electrode Charge (ppm) |
Element	Analytical Wavelength (A)				
B	2496.8 (second order)	4.6–>224	4.1	0.5–13	0.6
Pb	2833.07	2.8–~249	8.0	2.2–15	5
Cu	3273.96	4.5– 69	5.4	1.3–20	2
Co	3453.50	1.0– 42	4.1	0.5–19	2
Ni	3414.76	3.8– 105	4.0	0.9–11	1.5
Be	2348.61	0.5– 5.9	3.1	0.0–12	0.5
Cr	4254.35	4.4– 33	3.4	0.9–10	1.5
V	3185.40	8.4–~108	6.8	2.4–12	2
Sn	3034.12	1.0– 51	18.0	4.7–57	6
Mo	3170.35	<0.3– 30	5.0	1.7–12	0.2
Ge	2651.18	<0.7– 35	12.0	3.0–29	0.3

Table VII.   Statistical Results of the Spectrographic Method

		Coal		Ash	
Element	Analytical Wavelength (A)	Concentration Range (ppm)	Average Relative Standard Deviation (%)	Upper Determination Limit (ppm)	Detection Limit (ppm)
B	2497.7	11– 325	12	~3000[b]	85
Pb	2833.07	3.7– 370	9	~3700[b]	30
Cu	3273.96	4– 65	18	~500	50
Co	3405.1	1.8– 49	25	~1300[b]	10
Ni	3414.76	1.3– 60	24	~900	20
Be	3130.4	0.20– 4.5	14	~40	2.1
Cr	2677.2	<3.8– 50	23	~1000	45
V	3184.0	11– 133	30	1200[a]	60
Ge	2651.18	1.2– 117	16	430[a]	30
Ag	3280.7	0.16– 3.7	[d]	26[a]	~1
Zn	3345.0	7–>200	[e]	~4000	100
Zr	3392.0	8.4–>263	[c]	~3000[b]	50
Mn	2576.1	5.2–>400	19[e]	~5300[b]	100

[a] Highest concentration analyzed
[b] Estimated by extrapolation past 1000 ppm
[c] This figure is meaningless because the variation between duplicates was so great, probably because of the presence of zircon particles (see text).
[d] Too few concentrations found above minimum
[e] Too many concentrations found above maximum

sample containing a small particle of the oxide originating from a discrete mineral could give a discernibly higher result for that element only and still give accurate determinations of other elements. Data showing concentrations of trace elements in 25 coals have been given by Ruch et al. (10).

Summary

Two reliable optical emission spectroscopic methods have been developed for trace element analyses of high-temperature ash from coal samples. However, care must be taken in the ashing procedure to guard against contamination or loss of certain elements.

Literature Cited

1. Abernethy, R. F., Peterson, M. J., Gibson, F. H., U.S. Bur. Mines Rep. Invest. (1969), No. 7281.
2. Zubovic, P., Stadnichenko, T., Sheffey, N. B., U.S. Geol. Surv. Bull. (1961), No. 1117-A, A18–A20.
3. Ibid., (1964), No. 1117-B.
4. Ibid., (1966), No. 1117-C.
5. Zubovic, P., Sheffey, N. B., Stadnichenko, T., U.S. Geol. Surv. Bull. (1967), No. 1117-D.

6. Rao, P. D., *Alaska Univ. Miner. Ind. Res. Lab. Rep.* (1968), No. **15**.
7. Hunter, R. G., Headlee, A. J. W., *Anal. Chem.* (1950), **22**, 441.
8. Shimp, N. F., Leland, H. V., White, W. A., *Ill. State Geol. Surv. Environ. Geol. Notes* (1970), No. **32**, 5.
9. Gorsuch, T. T., *Analyst* (1959), **84**, 135.
10. Ruch, R. R., Gluskoter, H. J., Shimp, N. F., *Ill. State Geol. Surv. Environ. Geol. Notes* (1973), No. **61**, 5–6, 28–30.
11. Von Lehmden, D. J., LaFleur, P. D., Akland, G. G., "Abstracts of Papers," *166th Meeting, ACS Div of Fuel Chem.* (1973), **34**.
12. Gluskoter, H. J., *Fuel* (1965), **44**, 285.
13. Williams, D. E., Vlamis, J., *Soil Sci.* (1961), **92**, 161.
14. Scott, R. O., Mitchell, R. L., Purves, D., Voss, R. C., *Consult. Comm. Develop. Spectrochem. Work, Bull.* (1971), No. **2**, The Macaulay Institute for Soil Research, Aberdeen.
15. Mitchell, R. L., *Commonw. Bur. Soils, Harpenden, Tech. Commun.* (1964), No. **44a**.
16. Davidson, M. M., Mitchell, R. L., *J. Soc. Chem. Ind. London Trans., Part II* (1940), **59**, 213.
17. Gluskoter, H. J., Lindahl, P. C., *Science* (1973), **181**, 264.
18. Clark, M. C., Swaine, D. J., *Commonw. Sci. Ind. Res. Org., Tech. Commun.* (1962), No. **45**, 24–28.
19. Abernethy, R. F., Gibson, F. H., *U.S. Bur. Mines Inform. Cir.* (1963), No. **8163**, 13–33.

RECEIVED February 27, 1974.

4

Trace Elements in Coal Dust by Spark-Source Mass Spectrometry

A. G. SHARKEY, JR., T. KESSLER, and R. A. FRIEDEL

Spectro-Physics, Pittsburgh Energy Research Center, 4800 Forbes Ave., Pittsburgh, Penn. 15213

Spark-source mass spectrometry has been used to determine trace elements in 10 pairs of respirable-range mine dusts and prepared coal dusts. The object of the research is to obtain information for correlating the incidence of coal workers' pneumoconiosis with the composition of mine dusts. Samples representing eight coal seams in Pennsylvania, West Virginia, Virginia, and Utah were analyzed. For a majority of the sample pairs, several elements including Ag, Cd, Cu, Cr, Rb, Ca, Cl, P, and Br show higher values in the mine dust than the coal. A limited investigation of the organic material associated with the mine dusts indicates an additional series of highly saturated material not derived from the coal.

Coal workers' pneumoconiosis (CWP) is a major health hazard among miners in the Appalachian coal regions, disabling many and con‧tributing to the death of others. It is now widely accepted that CWP results from the inhalation of coal dust, but, since the incidence of the disease among miners appears to vary with the region or coal bed being mined, identification of the specific lung irritant(s) responsible may help to minimize or eliminate the hazard. However, information currently available on coal dust composition is insufficient to attempt any correlation with CWP incidence.

Lee summarized the panel discussion at a meeting of the Environmental Health Sciences Advisory Council which concentrated primarily on the etiology of CWP (1). One area of research recommended by the panel was what component(s) is associated with a specific rank of coal that results in increased incidence of CWP. Warden, in discussing the medical aspects of CWP, stated that "the fate of the particulates retained in the lung is related to the solubility of the particulate, as well as to its chemical activities, surface reactivity, and immunological factors" (2). Saffiotti and co-workers demonstrated in animal tests that polynuclear aromatic compounds, such as are prevalent in coal dust, when combined with inorganic material produce some of the most severe lung irritations (3). Langer and Selikoff showed that the chemical composition of asbestos fibers is altered when in contact with lung tissue because fluids associated with lung tissue are effective in leaching magnesium from the asbestos (4). The previously stated investigations indicate that both organic and inorganic constituents of the coal dust could have a bearing on the incidence of CWP.

Coal workers' pneumoconiosis can be defined simply as the accumulation of dust in the lung and the reaction of lung tissue to the presence of that dust. The term, respirable dust, has been given to dust particles less than 5 μm in size, although a few particles up to 10 μm have been observed in lungs. The respirable dust generally has a composition quite different from dust in the same size range, which is prepared from coal from the same mine. Considerable contamination of airborne dust can occur from other materials such as rock dust (calcium carbonate), slate, and organic contaminants. The mine dust and coal dust can also differ in shape factor, density, and other properties.

Very little data have been reported on the analysis of elements in whole coal and mine dusts in particular. Kessler, Sharkey, and Friedel analyzed trace elements in coal from mines in 10 coal seams located in Pennsylvania, West Virginia, Virginia, Colorado, and Utah (5). Sixty-four elements ranging in concentration from 0.01 to 41,000 ppm wt were determined. Several surveys published previously have provided data on the concentration of minor elements in ashes from coals rather than a direct determination on the whole coals or mine dusts. Previous investigations include studies by Headlee and Hunter (6), Nunn, Lovell, and Wright (7), Abernethy, Peterson, and Gibson (8), and others (9, 10, 11, 12).

The purpose of this investigation is to determine if there are any differences in composition between samples of respirable dust collected in coal mines and coal dust of respirable size and similar particle size distribution prepared from a representative coal sample from the same mine. This report hopes to provide survey analyses indicating the general concentration level of many trace elements in the respirable-range mine dust, including heavy metals of high current interest. More precise data can be obtained by atomic absorption and neutron activation or other techniques for certain trace elements showing high or hazardous concentrations in the mine dusts analyzed by spark-source mass spectrometry (SSMS). The determination of possible differences in the organic material associated with the same 10 sample pairs is also part of this program. High-resolution mass spectra are included to illustrate the data obtained.

Experimental Work

Sample Preparation. Representative samples for the analysis of the elements in the coal were obtained by the following procedure.

400 lbs of raw coal were obtained from each of the seams investigated. The raw coal sample was first crushed to $\frac{3}{8} \times 0$ in. The crushed coal was divided by successive riffling to obtain a final 25-lb sample which was then air dried and divided twice to obtain a 6.25-lb sample. The 6.25-lb sample ($\frac{3}{8} \times 0$ in.) was reduced to -200 mesh by grinding on a micro mill equipped with Stellite blades. Using a wind tunnel—Anderson Sampler system, the -200-mesh coal was separated into the following particle sizes: > 9.2, 5.5–9.2, 3.3–5.5, 2.0–3.3, 1.0–2.0, and < 1.0 μm.

Previous work has shown that the 3.3–5.5 μm coal fraction has a particle size distribution similar to that of respirable mine dust collected on personal samplers during mining operations (13). This fraction was used for the spark-source analyses. The coals used in this investigation are identified in Table I. The respirable dusts were obtained from personal sampler filters submitted to the Dust Group, Pittsburgh Technical Support Center, Federal Bureau of Mines. The samples were collected during actual mining operations.

Table I.   Identification of Seams

Sample No.	Rank[a]	Seam	Location
1	mvb	Lower Kittanning	Cambria Co., Pa.
2	hvAb	Powellton	Fayette Co., W. Va.
3	hvAb	Thick Freeport	Allegheny Co., Pa.
4	hvAb	Pittsburgh	Greene Co., Pa.
5	hvAb	Pittsburgh	Allegheny Co., Pa.
6	hvAb	Sunnyside	Carbon Co., Utah
7	lvb	Pocahontas No. 4	McDowell Co., W. Va.
8	lvb	Pocahontas No. 3	Buchanan Co., Va.
9	lvb	Pocahontas No. 3	Buchanan Co., Va.
10	lvb	Taggart	Wise Co., Va.

[a] mvb = medium volatile bituminous, hvAb = high volatile A bituminous, lvb = low volatile bituminous.

Instrumentation. The instrument used in this investigation was a commercial Mattauch–Herzog mass spectrometer equipped with photographic and electrical detection systems and an RF spark source. The resolution of the instrument was 1 part in 5,000. All trace elements were determined from mass spectra recorded on Ilford Q-2 photographic plates. Major elements were determined using the electrical detection system.

Electrode Preparation. Electrodes were prepared by mixing the samples with equal parts of pure graphite. To insure homogenous mixing and to determine the plate sensitivity, 50 ppm indium was added as an internal standard to the sample–graphite mixture. (In this paper, ppm are given in terms of weight.) The mixtures were pressed into

1	2	3	4	5	6	7	8	9	10	11	12	13	14	15	16	17
H																
ND																
Li	Be											B	C	N	O	F
100	100											100	ND	ND	ND	100
Na	Mg											Al	Si	P	S	Cl
100	100											100	100	100	100	100
K	Ca	Sc	Ti	V	Cr	Mn	Fe	Co	Ni	Cu	Zn	Ga	Ge	As	Se	Br
100	100	100	100	100	100	100	100	100	100	100	100	100	100	100	100	100
Rb	Sr	Y	Zr	Nb	Mo	Tc	Ru	Rh	Pd	Ag	Cd	In	Sn	Sb	Te	I
100	100	100	100	100	100	Nd	0	0	0	92	92	Standard	100	92	85	85
Cs	Ba	La	Hf	Ta	W	Re	Os	Ir	Pt	Au	Hg	Tl	Pb	Bi	Po	At
100	100	100	46	62	69	0	0	0	0	0	38	31	100	31	ND	ND
Fr	Ra	Ac														
ND	ND	ND														

Ce	Pr	Nd	Pm	Sm	Eu	Gd	Tb	Dy	Ho	Er	Tm	Yb	Lu
100	100	100	ND	100	100	85	85	85	77	77	0	62	38

Th	Pa	U
92	ND	92

Figure 1. Occurrence frequency of elements in 13 raw coals as determined by spark-source mass spectrometry. All quantities in %. ND = not determined. O = checked but not detected.

1	2	3	4	5	6	7	8	9	10	11	12	13	14	15	16	17
H																
ND																
Li	Be											B	C	N	O	F
4-163	0.4-3											1-230	ND	ND	ND	1-110
Na	Mg											Al	Si	P	S	Cl
100-1000	500-3500											3000-23,000	5000-41,000	6-310	700-10,000	10-1500
K	Ca	Sc	Ti	V	Cr	Mn	Fe	Co	Ni	Cu	Zn	Ga	Ge	As	Se	Br
300-6500	800-6100	3-30	200-1800	2-77	26-400	5-240	1400-12,000	1-90	3-60	3-180	3-80	0.3-10	0.03-1	1-10	0.04-0.3	1-23
Rb	Sr	Y	Zr	Nb	Mo	Tc	Ru	Rh	Pd	Ag	Cd	In	Sn	Sb	Te	I
1-150	17-1000	3-25	28-300	5-41	1-5	ND	<0.1	<0.1	<0.1	<0.01-3	<0.01-0.7	Standard	1-47	<0.1-2	<0.1-0.4	<0.1-4
Cs	Ba	La	Hf	Ta	W	Re	Os	Ir	Pt	Au	Hg	Tl	Pb	Bi	Po	At
0.2-9	20-1600	0.3-29	<0.3-4	<0.1-8	<0.1-0.4	<0.2	<0.2	<0.2	<0.3	<0.1	<0.3-0.5	<0.1-0.3	1-36	<0.1-0.2	ND	ND
Fr	Ra	Ac														
ND	ND	ND														

Ce	Pr	Nd	Pm	Sm	Eu	Gd	Tb	Dy	Ho	Er	Tm	Yb	Lu
1-30	1-8	4-36	ND	1-6	<0.1-0.4	<0.1-3	<0.1-2	<0.1-5	<0.1-0.4	<0.1-0.4	<0.1	<0.1-0.5	<0.1-0.3

Th	Pa	U
<0.1-5	ND	<0.1-1

Figure 2. Concentration range of elements in 13 raw coals analyzed by spark-source mass spectrometry. All quantities in ppm wt. ND = not determined.

electrodes in polyethylene slugs in a commercial isostatic die. The final electrode size was $1/16 \times 3/8$ in.

Interpretation of Spectra. For the analysis of trace elements, a series of graded photoplate exposures ranging from 1×10^{-12} to 3×10^{-7} coulomb was made. Using the same sparking parameters and series of graded exposures used for the samples, mass spectra were obtained for the standard, U. S. Geological Survey sample BCR-1 (Basalt rock). The concentrations of 63 elements have been determined in this sample by a variety of techniques, including SSMS (14). Determinations of the various elements present in the sample spectra were made by visual comparison with the spectra obtained from the BCR-1 standard using the following expression:

$$\text{Concentration of trace element} = \frac{E_{\text{standard}}}{E_{\text{sample}}} \times \frac{\text{known concentration of}}{\text{element in standard}}$$

Table II.   Partial Spark-Source Mass Spectrometry Analyses of

Seam[a]	Rank	Dust[b]	Coal[c]	Dust/Coal	Dust	Coal	Dust/Coal
			Ag			*Cd*	
1	mvb	0.2	0.02	10	39	0.2	195
2	hvAb	0.5	0.02	25	0.8	0.2	4
3	hvAb	2	0.03	67	0.3	0.7	0.4
4	hvAb	0.6	0.01	6	0.1	0.04	2.5
5	hvAb	0.2	0.04	5	3	0.07	43
6	hvAb	0.2	<0.02	>10	2	<0.1	>20
7	lvb	0.09	0.03	3	0.04	1.0	0.04
8	lvb	0.02	0.02	1	0.7	0.2	3.5
9	lvb	0.2	0.02	11	3	0.04	75
10	lvb	4	0.03	133	6	0.3	20
			Rb			*Ca*	
1	mvb	1	15	0.1	0.3	0.1	3
2	hvAb	88	11	8	0.4	0.2	2
3	hvAb	34	44	0.8	0.2	0.1	2
4	hvAb	20	20	1	1.8	0.3	6
5	hvAb	88	15	5.9	1.7	0.2	8.5
6	hvAb	15	9	1.7	2.3	0.09	11
7	lvb	5	1	5	0.5	0.2	2.5
8	lvb	44	10	4.4	1.4	0.2	7
9	lvb	88	12	7.3	1.1	0.1	11
10	lvb	88	150	0.6	0.9	0.1	9

[a] Identification of seam is given in Table I.
[b] Respirable dust
[c] 3.3–5.5 μm Anderson Sampler coal fraction

where E is the exposure producing the same blackening of the photo-plate for the same element isotope in both sample and standard spectra.

For example, if the 1×10^{-10} coulomb exposure for $^{98}Mo^{+1}$ for a sample produced the same blackening as did the 1×10^{-12} coulomb exposure for $^{98}Mo^{+1}$ in the BCR-1 standard that contained 3.8 ppm of molybdenum, then the concentration of molybdenum in the sample is:

$$\frac{1 \times 10^{-12}}{1 \times 10^{-10}} \times 3.8 \text{ ppm} = 0.038 \text{ ppm}$$

Major elements were determined by magnetic scanning of the mass range of interest and recording the signals electrically. Calculations of the concentrations of the major elements were made using the following equation:

mvb, hvAb, and lvBb Respirable Dusts and Coals (ppm wt)

Dust	Coal	Dust/Coal	Dust	Coal	Dust/Coal
	Cu			*Zr*	
22	6	3.6	70	30	2.3
18	6	3	37	28	1.3
20	18	1.1	37	28	1.3
560	10	56	3000	28	107
500	150	3	1800	300	6
45	180	0.3	370	30	12
60	15	4	300	25	12
18	3	6	99	30	3.3
22	18	12	99	30	3.3
180	180	1	740	99	7.5

Cl			*P*			*Br*		
600	34	18	620	47	13	230	23	10
340	25	14	470	47	10	33	2	17
17	100	0.2	62	230	0.3	45	18	2.5
100,000	150	667	12,000	160	75	60	0.2	300
6,500	8	812	20,000	310	65	18	0.1	180
170	10	17	310	310	1	30	0.4	75
210	40	5.3	1,600	6	266	4	0.3	13
1,000	50	20	2,100	19	111	23	18	1.3
1,000	1,000	1	1,900	23	83	18	5	3.6
3,100	6	517	1,900	23	83	23	0.2	115

$$\text{Concentration of} \atop \text{major element} = \frac{\text{peak height }_{\text{sample}}}{\text{peak height }_{\text{standard}}} \times {\text{known concentrations} \atop \text{of element in standard}}$$

The same isotope is used for both sample and standard.

High-resolution mass spectra of coal and mine dusts are also being obtained to determine if there are differences in the organic material associated with the two types of samples. The distribution of carbon–hydrogen is being investigated at present.

Results and Discussion

To provide background information for the data obtained in this investigation, a summary of the spark-source determinations on 13 coals from 10 seams is given in Figures 1 and 2. Figure 1 shows the occurrence frequency of the 64 elements analyzed, and Figure 2 gives the concentration ranges of the elements. Analyses for the individual elements were reported previously (5).

The elemental compositions of the respirable dust and 3.3–5.5 μm coal fraction from the 10 mines are similar. It has been established in several investigations that the error of measurement of any single element by the visual comparison method is less than a factor of two (15). For most trace elements the precision is ±25%. The precision was verified in a current series of analyses in which the concentration of trace elements as a function of particle size was determined (13). The only elements showing a consistent difference for a majority of the respirable dusts and corresponding coal fractions from the various mines are given in Table II. The elements showing a concentration difference greater than a factor of two are Ag, Cd, Zr, Rb, Ca, Cl, P, and Br. The majority of samples show higher concentrations of the specified elements in the respirable dust than in the corresponding 3.5–5.5 μm coal fraction. The average ratios of the concentration of the various elements in the dust compared with the coal are as follows: Ag, 27; Ce, 36; Cu, 9; Zr, 16; Rb, 4; Ca, 6; Cl, 200; P. 71; and Br, 72. It is interesting to note that high dust-to-coal ratios for both chlorine and bromine occur in samples from seams 4, 5, and 10.

Differences observed in the organic material in mine dust and coal from the same mine are illustrated in Figure 3. High-resolution mass spectral data for the carbon–hydrogen (C–H) distributions in less than 74-μm and 3.5–5.5 μm coal and coal dust are shown. Formulas were detected for compounds with C–H values included in the lines. For this particular coal, many of the C–H ratios approach 1, indicating the highly aromatic character of the organic material. An additional series of highly saturated material, not present in the coal, was detected in the mine dust.

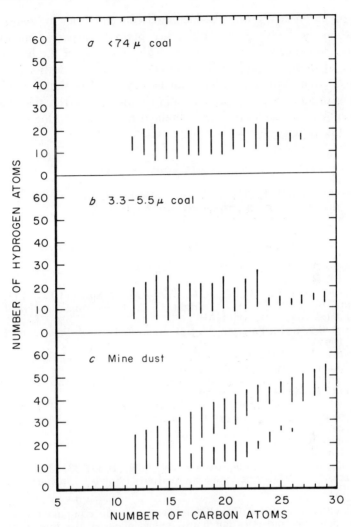

Figure 3. Organic material in mine dust and coal from the same mine

The source of this saturated material is not known, but possibly it is from hydraulic fluids used in underground engines.

Summary

This investigation has shown that for the majority of the 10 pairs of dust samples investigated, nine elements including Ag, Cd, Cu, Zr, Rb, Ca, Cl, P, and Br show much higher concentrations in the mine dusts collected during mining operation than in the corresponding respirable-range prepared coal dust. The increased concentrations are considerably more than the factor of two difference normally expected in duplicate runs in spark source survey analyses. The source(s) of the high concentrations of these specific elements is not known but additional data are required before any conclusions can be drawn. The significance of the highly saturated series of organic compounds detected in the mine dust is not known but data are being obtained for several additional dusts to verify this finding.

Literature Cited

1. Lee, Douglas H. K., *J. Occup. Med.* (1971) **13**, 183.
2. Warden, Henry F., Jr., *Mining Congr. J.* (1969) **55**, 78.
3. Saffiotti, U., Cefis, F., Kolb, L. H., Shubik, P., *J. Air Pollut. Contr. Ass.* (1965) **15**, 23.
4. Langer, A. M., Selikoff, L. J., *Proc. Int. Clean Air Congr., 2nd,* New York (1971).
5. Kessler, T., Sharkey, A. G., Jr., Friedel, R. A., *U.S. Bur, Mines* **RI 7714** (1973).
6. Headlee, A. J. W., Hunter, R. G., *W. Va. Geol. Surv.* (1955) **13A**, 36.
7. Nunn, R. C., Lovell, H. L., Wright, C. C., *Trans. Annu. Anthracite Conf. Lehigh Univ.* (1953) **11**, 51.
8. Abernethy, R. F., Peterson, M. J., Gibson, F. H., *U.S. Bur. Mines* **RI 7281** (1969).
9. Adams, H. F., "The Seams of the South Wales Coalfield," The Institute of Min. Eng. Lond. (1967), Monogr.
10. Zubovic, P., Stadnichenko, T., Sheffey, N. B., *Geol. Surv. Bull.* **1117-A, 1117-B, 1117-C, 1117-D** (1961).
11. Bethel, F. V., "Progress Review No. 55," *J. Inst. Fuel* (1963) **36**, 478.
12. Zubovic, P., Stadnichenko, T., Sheffey, N. B., *Geol. Surv. Prof. Paper* **424-D** (1961).
13. Kessler, T., Sharkey, A. G., Jr., Friedel, R. A., *U.S. Bur. Mines* **TPR 42** (1971).
14. Flanagan, F. J., *Geochim. Cosmochim. Acta* (1969) **33**, 81.
15. Ahern, A. J., "Mass Spectrometric Analysis of Solids," Elsevier, Amsterdam, 1966.

RECEIVED February 27, 1974. Reference to specific equipment is made for identification only and does not imply endorsement by the Bureau of Mines.

Major and Minor Constituents in Siliceous Materials by Atomic Absorption Spectroscopy

RICHARD B. MUTER and LARRY L. NICE

Coal Research Bureau, West Virginia University, Morgantown, W. Va. 26506

Siliceous materials—Si, Al, Fe, Ti, Ca, Mg, Na, K, Mn, Ni, Ba, Ag, Au, Ca, Cr, Cu, Ga, In, Mo, Sb and Zn—may be analyzed by a lithium tetraborate fusion–acid dissolution technique using atomic absorption spectroscopy. Mercury, tin, and lead volatilize by this technique, and gold and silver in concentrations above 0.5 wt% cannot be held in solution. Coal ash is preconcentrated prior to analysis, and there is possible silica interference. Analytical results, where possible, are compared statistically with other reported values.

Increasing national concern over the ecological and environmental effects of coal combustion coupled with the desire to become more self sufficient in mineral production led the Coal Research Bureau at West Virginia University to examine the major and minor constituents in coal ash. Because of the need for accurate results at the low trace element concentrations, it was felt that atomic absorption spectroscopy could provide a rapid and routine method for analytical determinations.

The introduction of atomic absorption spectroscopy has resulted in major advances in the rapid analysis of many elements. Initially, atomic absorption was applied only to aqueous systems or to materials that could be readily solubilized. There are methods to analyze major elements in such complex materials as silicates and vitreous siliceous coal ashes (1–5). More recently, lithium metaborate has been reported to be a good fluxing agent (6) and has also been used in conjunction with atomic absorption analysis in silicate analysis (7). This paper describes a lithium tetraborate–atomic absorption analytical technique which is being used to analyze coal ash.

While the amount of ash in coal varies from rank to rank, the carbon content is relatively high, and at least one preconcentration step, ashing the sample, is required for the lithium tetraborate technique. Only two of the available ashing techniques require minimum sample handling— high temperature ashing (HTA) and low temperature ashing (LTA). High temperature ashing is the simplest to perform as it normally requires that the coal be heated for 2 hrs at 800°C. Unfortunately some of the more volatile trace elements such as mercury, tin, and lead are lost by this technique. In low temperature ashing, the coal is heated in an atomic oxygen atmosphere in which the reactive oxygen combines with carbon at temperatures around 100°–150°C. This temperature range is low enough that the mineral matter in coal is substantially unaltered by ashing, but ashing time is increased from 2 hrs to 24 hrs or longer. In addition it has been reported that some of the more volatile trace elements may also be lost by LTA (8). Because a rapid routine method for coal mineral matter analysis was needed, it was felt that high temperature ashing followed by lithium tetraborate fusion would produce a sample suitable for atomic absorption analysis.

Table I. Accuracy and Precision Data for

Element	Standard	Weight % Known Value	Weight % A.A. Value
Si	G-2	32.34	30.45
Al	G-2	8.12	7.45
Fe	G-2	1.93	2.14
Ti	BCR-1	1.34	1.29
Ca	W-1	7.79	7.70
Mg	W-1	3.98	4.09
Na	W-1	1.56	1.61
K	W-1	0.55	0.54
Mn	G-2	0.023	0.024
Ni	Si-Al	1.0	0.969
Ba	Synthetic 3[a]	2.45	2.333
Ag	Synthetic 1[a]	0.63	0.609
Au	Noble-G	0.50	0.515
Co	Synthetic 4[a]	8.48	8.49
Cr	Si–Al	1.0	1.031
Cu	Synthetic 1[a]	0.63	0.626
Ga	Si–Al	1.0	1.00
Hg	Synthetic 1[a]	1.57	1.599
In	Noble-G	0.50	0.41
Mo	Si–Al	1.0	0.90
Pb	Synthetic 1[a]	1.26	1.26
Sb	Synthetic 2[a]	4.34	4.30
Sn	Synthetic 2[a]	8.86	8.53
Zn	Si–Al	1.0	0.966

[a] Synthetic standards were not fused but were prepared from aqueous standards with

Technique and Instrumentation

Apparatus. A Perkin-Elmer model 303 atomic absorption spectrometer equipped with a DCR-1 readout accessory and a strip chart recorder was used for all determinations. A Boling burner was used for all determinations made in the air–acetylene flame except for copper where a single-slot, high-solids burner was used. The nitrous oxide burner was used for refractory elements. Burner and instrument settings used were those recommended by the manufacturer's handbook.

Preparation of Standards. Standards for ash analysis were prepared from commercially available pure salts in aqueous solution with appropriate acids addition where necessary to match acid concentrations in the samples as well as to hold materials in solution. Master standard solutions were prepared so that serial dilutions for the construction of working curves were possible. A constant amount of silicon and aluminum (equivalent to 20% Si–5% Al) interference solution was added to each set of standards along with lithium tetraborate to carefully match

Standards Used to Check the Working Curves

Absolute Error (%)	Relative Error (%)	Standard Deviation (%)	Coefficient of Variation (%)
1.89	5.84	0.77	2.52
0.67	8.25	0.19	2.55
0.21	10.88	0.05	2.33
0.05	3.73	0.085	6.58
0.09	1.16	0.13	1.69
0.11	2.76	0.12	2.93
0.05	3.20	0.14	8.70
0.01	1.82	0.03	5.56
0.001	4.35	0.004	16.67
0.031	3.10	0.020	2.06
0.117	4.78	0.034	1.46
0.021	3.33	0.027	4.43
0.015	3.00	0.012	2.33
0.01	0.12	0.07	0.82
0.031	3.10	0.058	5.62
0.004	0.63	0.012	1.92
0	0	0	0
0.029	1.85	0.145	9.07
0.09	18.00	0.018	4.39
0.10	10.00	0.03	3.33
0	0	0	0
0.04	0.92	0.10	2.33
0.33	3.80	0.19	2.23
0.034	3.40	0.075	7.76

appropriate acid, lithium tetraborate, and interference solution additions.

Table II.   Comparison of U.S.G.S. Trace Element Concentrations

	G-2		GSP-1		AGV-1	
	Fusion	Reported Value	Fusion	Reported Value	Fusion	Reported Value
Cr	100	7	80	12.5	130	12.2
Mn	240	260	290	331	650	763
Ni	400	5.1	350	12.5	480	18.5

sample and standard solution characteristics. Standards with adjusted silicon–aluminum and lithium tetraborate concentrations were prepared in the event that sample dilution was needed (*i.e.*, calcium determination) for analysis.

Sample Preparation. The following method was used to prepare solid samples for analysis.

One tenth g of sample is added to a plastic vial containing 1 g of preweighed lithium tetraborate. The vial is hand shaken to mix the material, and the contents are poured into a graphite crucible. The material is fused at 950°C for 15 min in a muffle furnace. The resulting bead is removed from the furnace and can either be stored in the original vial or immediately solubilized. The bead is transferred to a Teflon beaker containing 5 ml of 3N HCl, 2 ml of 2N HNO$_3$ and 10 ml of water. Teflon is used to eliminate sodium contamination. The material is then boiled until completely dissolved and immediately filtered into a 50 ml volumetric flask. Hot filtration is required to prevent solid materials from crystallizing out of solution before dilution and to remove carbon particles that result from fusion in graphite crucibles. The sample is then diluted to volume, shaken, and then further diluted as required to bring the element concentration to within range of the working curve of interest. For the determination of calcium, magnesium, and barium, lanthanum chloride is added as a releasing agent to achieve a final lanthanum concentration of 1% as recommended by the spectrometer manufacturer.

Results and Discussion

Although detectable concentrations for several elements could be found after fusion, it is felt that the volatility of mercury and possibly lead and tin would make their determination by lithium tetraborate fusion questionable. Table I shows the elements selected for analysis and the accuracy and precision data for the standards used to check the fusion method. Each standard in Table I was of known composition and siliceous in nature. The standards were separately prepared 10 times so that a statistical evaluation of the results could be made. The standards used were USGS Standards G-2, W-1, BCR-1, commercially prepared silica–alumina based standards, and unfused synthetic standards prepared by the Coal Research Bureau (*9, 10, 11, 12*). The synthetic standards were used because no commercially prepared standard having

and Concentration Found by Li$_2$B$_4$O$_7$ Fusion (ppm)

PCC-1		DTS-1		BCR-1	
Fusion	Reported Value	Fusion	Reported Value	Fusion	Reported Value
2730	2730	3750	4000	130	17.6
900	959	960	969	1360	1406
2150	2339	2000	2269	280	15.8

detectable concentrations of Ba, Ag, Co, Cu, Hg, Pb, Sb, and Sn were available. Salts of these eight elements were separately fused to determine if any difficulty might be expected in analyzing them by this technique. As expected, elements such as mercury, tin, and to a lesser extent, lead volatilized during fusion. Silver in high concentrations (above 0.5%) could not be solubilized. However, in low concentrations, both silver and gold could be fused and held in solution.

Relative error values for the elements ranged from zero to a high of 18% with most values being 5% or less. The 18% relative error was obtained for indium and is attributable to the low concentration of this element in the solution analyzed. Moreover, the indium values were obtained on the lower end of the working curve where the sensitivity is greatly reduced. Standard deviations and coefficients of variation for the elements of interest are at acceptable levels (less than 1% standard deviation and around 5% coefficient of variation) for this technique. Again it should be pointed out that the original purpose of the subject method was to develop a rapid routine analysis for the major and minor constituents in coal ash and related materials without the necessity of several preconcentration steps, solvent extraction techniques, or pH adjustments.

The application of the lithium tetraborate fusion technique to the analysis of siliceous ashes has resulted in over 10,000 elemental determinations. While detectable gold and silver concentrations have been found, the results are near the detection limits for those two elements.

Using the lithium tetraborate fusion technique, trace element concentrations were examined. Table II shows a comparison of results obtained for a few trace elements with other reported values (*12*). As can be seen from the table, results at higher concentrations are in fair agreement with the reported values for the three elements examined. Manganese compares favorably at all concentrations while chromium and nickel in the higher 2000 ppm range show fair agreement. Elements such as antimony, tantalum, zirconium, gold, and silver were too low in concentration to be determined by this technique.

In order to monitor the self-consistency of this technique, a raw (as received) sample was analyzed. Another portion of the same sample

Table III.   Comparison of Raw Sample and Calculated Materials Balance for Raw Sample Analysis

Element	Analyzed Raw Sample Composition (wt %)	Calculated Raw Sample Composition (wt %)
Al	11.72	11.03
Cr	0.065	0.059
Fe	3.09	3.21
Mg	0.98	0.92
Mn	0.14	0.11
Si	24.00	22.88

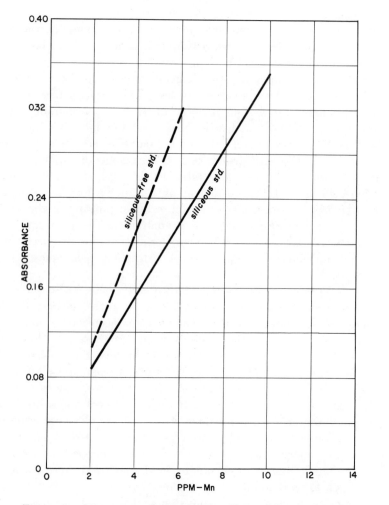

Figure 1.   Manganese determination illustrating silicon inter-ference

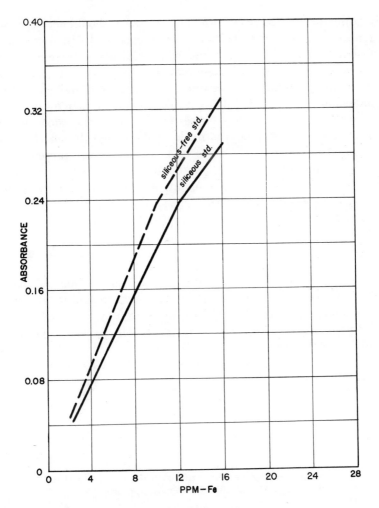

Figure 2.   Iron determination illustrating silicon interference

was separated according to screen size, and these sizes were also analyzed. Materials balance calculations were performed on the screened fractions. Chemical analysis and screen fraction percentages of each portion were used to back calculate a theoretical raw sample content and were then compared with the actual raw sample. Table III shows these results and indicates that the data are self-consistent.

Although prior research has established the interference effect of aluminum, previous in-house studies on the atomic absorption analysis of silicate materials indicated that silicon may also have an interference effect. This interference is probably caused by compound formation in the flame. The effect of silicon on the determination of all of the elements analyzed by the air–acetylene flame (interferences should not occur when using nitrous oxide–acetylene flame) was examined. In some cases a depression in absorbance was found when silicon was present, and in a few cases enhancement was observed. For example, depressions were found for iron, calcium, magnesium and manganese and enhancement for potassium and gallium (at higher concentrations). Figures 1 and 2 illustrate the types of working curves obtained and the effect of silicon on their slope. No attempt was made to determine the effect of interferences other than silicon because trace elements were present in such relatively minor quantities that interferences from them were not considered a problem. Interferences of this type can easily be removed by matching the concentration of silicon and aluminum in the standards to the concentrations generally expected in the unknown.

Conclusions

Lithium tetraborate has been found to be an excellent fusion agent enabling complete dissolution of silicate materials in acid for the analysis of major and minor constituents in coal. Carefully prepared standards matching the approximate concentrations of both the silica and alumina present in unknown samples permit determinations to be made with precision and accuracy. This method is currently being used to analyze coal ash and related materials.

Literature Cited

1. Muter, R. B., Cockrell, C. F., "The Analysis of Sodium, Potassium, Calcium and Magnesium in Siliceous Coal Ash and Related Materials," *Appl. Spectros.* (1969) **23** (5), 493.
2. Galle, O. Karmie, "Routine Determination of Major Constituents in Geologic Samples by Atomic Absorption," *Appl. Spectros.* (1968) **22** (5), 404.
3. Langmyhr, F. J., Paus, P. E., "Hydrofluoric Acid Decomposition Atomic Absorption Analysis of Inorganic Siliceous Materials," *At. Absorption Newslett.* (1968) **7** (6), 103.
4. Van Loon, J. C., "Determination of Aluminum in High Silica Materials," *At. Absorption Newslett.* (1968) **7** (1), 3.
5. Yule, John W., Swanson, Glenda A., "A Rapid Method for Decomposition and The Analysis of Silicates and Carbonates by Atomic Absorption Spectroscopy," *At. Absorption Newslett.* (1969) **8** (2), 30.
6. Ingamells, C. O., "Lithium Metaborate Flux in Silicate Analysis," *Anal. Chim. Acta* (1970) **52**, 323.
7. Medlin, J. H., *et al.*, "Atomic Absorption Analysis of Silicates Employing LiBO₂ Fusion," *At. Absorption Newslett.* (1969) **8** (2), 25.
8. Ruch, R. R., Gluskoter, H. J., Shimp, N. F., "Occurrence and Distribution of Potentially Volatile Trace Elements in Coal: An Interim Report," *Ill. State Geol. Surv. Environ. Geol. Notes* (1973) **61**.
9. Shapiro, Leonard, Brannoch, W. W., "Rapid Analysis of Silicate Rocks," *U.S. Geol. Surv. Bull.* (1956) **1036-C**.

10. Flanagan, F. J., "U. S. Geological Survey Standards—II. First Compilation of Data for the New U.S.G.S. Rocks," *Geochim. Cosmochim. Acta* (1969) **33**, 81.
11. Flanagan, F. J., "1972 Values For International Geo-chemical Reference Samples," *Geochim. Cosmochim. Acta* (1973) **37**, 1189.
12. Abbey, Sydney, "'Standard Samples' of Silicate Rocks and Minerals—A Review and Compilation," *Geol. Surv. Can.* (1972) paper **72-30**.

RECEIVED June 6, 1974. Use of trade names does not imply endorsement by the Coal Research Bureau.

6

X-Ray Fluorescence Analysis of Whole Coal

JOHN K. KUHN, WILLIAM F. HARFST, and NEIL F. SHIMP

Illinois State Geological Survey, Urbana, Ill. 61801

X-ray fluorescence analysis proved to be rapid, simple, and reasonably accurate for determining the concentration of 21 minor and trace elements in whole coal. Development of the method was facilitated by the availability of a set of samples that had been analyzed during the investigation of over 100 coals for trace and minor elements by optical emission spectroscopy, atomic absorption spectroscopy, neutron activation, and wet chemical techniques. Although major elements in coal, carbon, hydrogen, oxygen, and nitrogen, cannot be analyzed by x-ray fluorescence, most other elements at levels greater than a few parts per million are readily determined. Tables present data supporting the conclusion that x-ray fluorescence may be the best method for analyzing large numbers of coal samples.

Recent interest in the trace element content of coal has increased the need for rapid and accurate analytical methods for their determination. Because x-ray fluorescence analysis has demonstrated its usefulness in determining major, minor, and trace elements in numerous other types of materials, it was felt that this method could be extended to trace element determinations in whole coal. In the past, such analyses were seriously hampered by the lack of standard samples. However, research being conducted in our laboratories under the sponsorship of the U. S. Environmental Protection Agency produced a large number of coal samples for which trace elements had been determined by two or more independent analytical procedures, for example, optical emission, neutron activation, atomic absorption, and wet chemical methods. These coals were used as standards to develop an x-ray fluorescence method that would determine many trace and minor elements in pressed whole coal samples.

The instrument used in this project was a Phillips manual vacuum x-ray fluorescence spectrometer. All analyses were made in the Analytical Chemistry Laboratories of the Illinois State Geological Survey.

Preliminary Investigation of Major and Minor Elements in Whole Coal and Coal Ash

Two different types of materials, coal ash and whole coal, were analyzed, and sample preparation was varied accordingly.

Whole coal was ground with a binder (10 wt %) and pressed into a disk, which was used as the analytical sample. The binder was a commercial product, Somar Mix, and the sample was ground in a No. 6 Wig-L-Bug for 3 min. Pellets 1⅛ in. in diameter were then formed at 40,000 psi in a die designed for that purpose. Sample preparation techniques are given in detail in a previous publication (1).

Two g of coal made a disk that was infinitely thick, *i.e.*, no x-rays penetrated the sample, for soft x-rays emitted by light elements such as magnesium, silicon, aluminum, and calcium. However, for elements heavier than bromine, the sample weight had to be increased to attain infinite sample thickness.

The use of x-ray fluorescence was originally intended to obtain information about the major element matrix of coal ashes that were to be analyzed for trace elements by optical emission spectroscopy. Both low-temperature ($<150°C$) and high-temperature ($450°C$) coal ashes, prepared as described by Ruch *et al.* (1), were analyzed, and the method of Rose *et al.* (2) was adapted to determine the major and minor elements (Si, Ti, Al, Fe, Mg, Ca, K, and V). The instrumental parameters used for these elements are given in Table I.

To assess the validity of this procedure, a series of coal ashes analyzed by the British Coal Utilization Research Association (BCURA) were again analyzed with the two types of ash prepared in our laboratories. Calibrations for these analyses were prepared from U. S. Geological Survey and National Bureau of Standards rock standards, 1B, G1, W1, and Nos. 78, 79, 88. The values determined for the BCURA coal ashes agreed excellently with results obtained at BCURA by Dixon *et al.* (3). Standard deviations were calculated for the duplicate coal ash determinations: Si—0.07%, Ti—0.01%, Al—0.06%, Fe—0.05%, Mg—0.02%, Ca—0.03%, K—0.01%, P—0.01%, and V—1.3 ppm. These deviations are comparable with Class A wet silicate analyses and indicate a high degree of precision.

Because of these encouraging results and previous work on brown coals by Sweatman *et al.* (4) and Kiss (5), which indicated that major and minor elements could be determined in whole coal, a series of 25 coals was prepared for x-ray fluorescence analysis. For each coal, a low-temperature ash, a high-temperature ash, and the whole coal itself

Table I.   Spectrometer Parameters ($K\alpha$ x-ray)

Element	2θ Angle	Background 2θ	Crystal	X-Ray Tube	PHA Volts Base	PHA Volts Window
Si	108.01	111.01	EDDT	Cr	7	17
Al	142.42	145.95	EDDT	Cr	5	17
Ti	86.12	89.12	LiF	Cr[a]	5	18
Fe	57.51	60.51	LiF	Cr[a]	5	25
Ca	44.85	47.95	EDDT	Cr	14	30
K	50.32	53.90	EDDT	Cr	14	21
Mg	136.69	139.69	ADP	Cr	4	8
V	76.93	80.93	LiF	Cr[a]	5	16
S	75.24	78.38	EDDT	Cr	12	18
Cl	64.94	67.94	EDDT	Cr	11	19
P	110.99	113.99	Ge	Cr	9	15
Ni	48.66	50.36	LiF	Cr[a]	10	27
Cu	45.02	49.67	LiF	Cr[a]	11	28
Zn	41.79	44.25	LiF	Cr[a]	10	22
Pb[b]	28.24	31.24	LiF	Cr[a]	22	28
Br	29.97	35.12	LiF	Cr[a]	25	23
As	34.00	37.00	LiF	Cr[a]	24	23
Co	52.79	53.79	LiF	W	13	16
Mn	62.97	63.97	LiF	W	8	12
Mo	20.33	19.83 20.83	LiF	W	36	40
Cr	69.35	68.53	LiF	W	7	15

[a] A tungsten tube is used on these elements when already in place.
[b] $L\beta_1$ x-ray.

were prepared by the indicated procedures. When all values were converted to the whole coal basis, the agreement among the three types of coal materials was excellent (Table II), indicating that the simpler and more rapid whole coal technique is acceptable for determining major and minor elements.

Determination of Trace Elements in Whole Coal

Trace element determinations on whole coal have been severely handicapped by the lack of analyzed standards. Because of this it was necessary to prepare calibration curves from samples analyzed in our laboratories by independent methods. The accuracy of the x-ray fluorescence method is, therefore, dependent on the accuracy of the methods used to analyze the calibrating standards. It was too difficult to prepare standards by uniformly adding known quantities of trace elements to ground whole coal.

The light coal matrix of carbon, hydrogen, and oxygen and the relatively slight variation of heavier trace elements permit their determination with minimum interferences. The same whole coal procedures previously

Table II.   Mean Absolute Variation Between Values Determined From 25 Raw Coals and Their Ashes

Element	Mean Difference (%)	Maximum Difference (%)
Si	0.10	0.24
Al	0.08	0.12
Ti	0.012	0.030
Fe	0.10	0.17
Ca	0.04	0.12
K	0.02	0.04
P	0.002	0.005
Mg	0.010	0.015

described were used, and the trace and minor elements, P, V, Cr, Mn, Co, Ni, Cu, Zn, As, Br, Mo, and Pb, were determined directly in 50 whole-coals.

The accuracy of the x-ray fluorescence method was evaluated by calculating, from the 50 whole coals analyzed, the mean variation of each element from its mean concentration, determined by the other independent methods previously mentioned and listed in Table III. Detection

Table III.   Comparative Accuracy for Whole Coal and Limits of Detection Based on 50 Samples

Element	Accuracy	Limit of Detection
	per cent	per cent
Al	±0.08	0.012
Si	±0.10	0.016
S	±0.04	0.003
Cl	±0.01	0.0015
K	±0.02	0,003
Ca	±0.04	0.0005
Mg	±0.010	0.015
Fe	±0.10	0.005
	ppm	ppm
Ti	±6.3	7.5
V	±3.1	2.5
Ni	±1.9	3.5
Cu	±2.5	1.0
Zn	±23.0	2.0
As	±4.3	3.2
Pb	±7.7	1.8
Br	±1.0	0.5
P	±15.0	15.0
Co	±1.3	2.5
Mn	±3.4	4.5
Cr	±2.1	1.5
Mo	±5.2	5.0

limits, three standard deviations above background, for each element also are given in Table III.

The relative errors for all elements determined are given in Table IV. For completeness, data on minor elements in whole coal also are included in the trace element tables. These data indicate the precision obtained for the x-ray fluorescence analysis on replicates of 15 samples of whole coal ground to −325 mesh.

Table IV.   Deviation on −325 Mesh on 15 Samples of Whole Coal

Element	Standard Deviation	Relative Deviation (%)
	per cent	
Al	0.02	1.77
Si	0.05	1.96
S	0.01	0.532
Cl	0.003	1.13
K	0.004	2.26
Ca	0.005	1.65
Mg	0.002	3.88
Fe	0.02	1.26
	ppm	
Ti	4.16	0.564
V	1.58	3.84
Ni	1.12	4.29
Cu	0.75	3.92
Zn	3.61	1.37
As	0.94	2.49
Pb	1.53	2.29
Br	0.39	2.11
P	3.41	10.92
Co	0.43	4.79
Mn	4.14	7.53
Cr	1.14	4.35
Mo	3.11	23.9

X-Ray Matrix Corrections for Analysis of Whole Coal

Because of the lack of standards, variations in analyses made by other methods, and errors caused by coal sampling problems, it was difficult to evaluate the need for x-ray matrix corrections and to select the best method for applying them. However, corrections were necessary because some elements in whole coal such as iron, silicon, and sulfur may vary considerably. For these elements, corrections were applied indiscriminately to all samples, because it was impossible to determine the point at which matrix variations required a correction greater than the accuracy limits of the method. We elected to use the minimum number of corrections compatible with reasonably accurate results. Therefore, the elements Mg, Al, Si, P, S, Cl, K, and Mo were left uncorrected. While these determinations probably could be improved (6), they were shown to be adequate for our purposes (Table II). The titanium and vanadium values were corrected by using the variations in the iron content of the whole coal.

The method used for correcting the other elements for matrix variations was proposed by Sweatman *et al.* (4). Total mass absorption was determined by measuring the attenuation of the radiation in question by a thin layer of the sample to be analyzed. The mass absorption coefficient M was calculated by $M = A/W$ (ln Cs/Cx); where A = area of sample (cm^2), W = weight of sample (g), Cs = intensity of the standard (counts/sec), and Cx = intensity of the standard (counts/sec) attenuated by the thin layer coal elements determined. Using these coefficients, a corrected value was obtained for the elements determined, even when matrix variations were considerable. Great care was taken to press the coals to a uniform thickness so that the mass absorption coefficient was affected only by density (for which compensation was made) and matrix considerations.

Effect of Coal Particle Size on Analytical Precision of Trace Elements

Our results indicated that coals ground to −60 mesh failed to yield a consistently acceptable precision for most trace element determinations. Therefore, it was necessary to evaluate the errors associated with trace element determinations in coals ground to various particle sizes.

Nine coals, representing a range of trace element concentrations, were carefully ground to pass screens of various mesh sizes (Table V). Duplicate 2-g coal samples for each mesh size were weighed and then ground 3 min in a No. 6 Wig-L-Bug to further reduce particle size. The final grinding eliminated, as nearly as possible, any variation in the pressed coal disks, which were subsequently prepared for analysis (1). More than 1000 individual determinations were made in this study.

Table V gives the combined means of the differences between duplicate trace element determinations for each coal particle size analyzed. Both the means of the absolute differences (in ppm) and the means of

Table V. Mean Error for All Elements at Various Coal Particle Sizes

Mesh Size (M)	ppm	Error of Mean Element Concentration (%)
−60	±3.05	8.47
−100	±2.11	6.38
−200	±1.26	4.28
−325	±1.12	2.62
−400	±1.02	1.56
<400	±0.93	1.40

the relative differences (absolute difference expressed as a percentage of the concentration) are given. The results show a progressive improvement in precision with decreasing coal particle size.

The ranges of relative differences between duplicate analyses for several trace elements, each in three mesh sizes, are given in Table VI. With the exception of bromine, the ranges are narrower for the −200 and −325 sizes than they are for the −60-mesh coal.

Table VI.   Range of Relative Errors for Three Particle Sizes of Whole Coal (%)

Element	−60 Mesh	−200 Mesh	−325 Mesh
V	0.0—10.0	0.3— 5.0	0.3—4.0
P	2.0—18.0	2.0—10.0	1.5—7.5
Ni	1.5—25.0	0.0—20.0	1.5—8.0
Cu	0.8—20.0	0.2— 1.0	0.2—1.0
Zn	1.2—25.0	1.2—12.0	0.1—6.5
Pb	0.4—23.0	1.2— 9.5	0.4—5.0
As	0.1— 6.0	0.1— 4.0	0.0—1.5
Br	0.0— 4.0	0.0— 3.5	0.0—3.0

Progressive reduction in coal particle size from −60 to −400 mesh improved the precision for all elements except bromine. The combined mean relative error for all elements was reduced below 5% for coal ground to −200 mesh.

These data indicate that, for most purposes, acceptable precision can be obtained when −200-mesh coal samples are used. Further improvement is achieved by grinding the samples to −325 mesh, but this is unnecessary except for analyses that are to be used as standard values or for other special purposes. Variations in the original field sampling of coal would probably negate any improvements in precision that might be gained from grinding below −325 mesh. Although this study applies directly to x-ray fluorescence analysis of whole coal, it should also apply to any method in which a limited sample (∼3 g or less) is taken for analysis.

Discussion and Conclusions

Results of analysis of whole coal samples by x-ray fluorescence agreed well with values determined by several other independent methods (Table IV). Subsequent analyses of more than 100 coals have supported this conclusion, and the figures will be published in a forthcoming *Environmental Geology Note* by the Illinois State Geological Survey. Some variations among the methods occurred when concentrations of trace elements were high, especially for the more coarsely ground coals. Because this was true with the other methods investigated, as well as with the x-ray fluorescence method, we felt the variations resulted from sampling errors caused by discrete mineral particles, such as pyrite and sphalerite, in whole coal. Geologists at the Illinois State Geological Survey have confirmed the presence of these discrete particles with the scanning electron microscope.

It is apparent from Table IV that trace elements determined by the x-ray fluorescence method are limited to those occurring in whole coals at concentrations of at least a few parts per million. Elements such as selenium, mercury, and antimony, which are generally present in whole coal at levels below 1 ppm, cannot be determined by this method. The major elements in coal, hydrogen, carbon, oxygen, and nitrogen, cannot be determined by x-ray fluorescence, but this should not inhibit the use of the method for trace and minor element determinations.

Our results indicate that x-ray fluorescence is highly useful for rapid and reasonably accurate analyses of whole coal for trace elements. Because of the speed and simplicity of the method, it is highly adaptable to large-scale surveys of coal resources. A suite of 24 samples can be analyzed for 21 elements in 3 days by manual instrumentation. While this simple procedure can not be used to determine certain elements, the time-saving factor over other methods (40 or 50 to 1 in the case of bromine by neutron activation) without loss of accuracy may well make x-ray fluorescence the method of choice for many elements. Improved equipment, such as nondispersive systems and automation, could extend the application of x-ray analysis to a dominant position for determining trace elements in whole coal.

Literature Cited

1. Ruch, R. R., Gluskoter, H. J., Shimp, N. F., *Ill. State Geol. Survey Notes* (1973) **61**, 81 pp.
2. Rose, H. J., Alder, I., Flanagan, F. J., *U.S. Geol. Survey Prof. Papers* (1962) **450-B**, 80–82.
3. Dixon, K., Edwards, A. H., James, R. G., *Fuel* (1964) **43**, 331–347.
4. Sweatman, T. R., Norrish, K., Durie, R. A., *CSIRO* (1963) Misc. Rept. **177**, 30 pp.
5. Kiss, L. T., *Anal. Chem.* (1966) **38**, 1713.
6. Berman, M., Ergun, S., *Bureau Mines Rept. Invest.* (1968) **7124**, 20 pp.

RECEIVED January 21, 1974

7

Trace Impurities in Fuels by Isotope Dilution Mass Spectrometry

J. A. CARTER, R. L. WALKER, and J. R. SITES

Oak Ridge National Laboratory, Post Office Box Y, Oak Ridge Tenn. 37830

Spark source (SSMS) and thermal emission (TEMS) mass spectrometry are used to determine ppb to ppm quantities of elements in energy sources such as coal, fuel oil, and gasoline. Toxic metals—cadmium, mercury, lead, and zinc—may be determined by SSMS with an estimated precision of ±5%, and metals which ionize thermally may be determined by TEMS with an estimated precision of ±1% using the isotope dilution technique. An environmental study of the trace element balance from a coal-fired steam plant was done by SSMS using isotope dilution to determine the toxic metals and a general scan technique for 15 other elements using chemically determined iron as an internal standard. In addition, isotope dilution procedures for the analysis of lead in gasoline and uranium in coal and fly ash by TEMS are presented.

Elements considered toxic to living organisms are present in concentrations from low ppb to high ppm in coal and other fuels used as energy sources. Since large central power stations consume over half of the coal being used, the concentrated quantities of such potentially harmful elements as mercury, cadmium, lead, zinc, and others are appreciable. In the USA, for example, over 500 million tons of coal are consumed annually, so any element present in coal at the 1 ppm level generates 500 tons of waste. Many of these elements are concentrated in the particulate fly ash or in the bottom slag. Efficient electrostatic precipitators, however, prevent most of the fly ash from dispersing into the atmosphere. The fly ash and the bottom slag then become storage and containment problems.

This research effort has demonstrated the capability of spark source and thermal emission mass spectrometry for determining the fate of trace elements in coal-fired, central power plants. Additionally, isotope dilution methods for the analyses of lead, cadmium, and mercury in gasolines and other petroleum fuels have been developed and used for referee and evaluation purposes.

Experimental Procedure

Spark Source Mass Spectrometry. The spark source mass spectrometer (SSMS) used in this research was a commercial Mattauch–Herzog double focusing instrument. A schematic representation of it is shown in Figure 1. In a SSMS analysis, an ion beam of the substance being investigated is produced in a vacuum by igniting a spark between two conductors with a pulsed, high frequency potential of 50 kV. During this process, the electrode substance is evaporated and ionized. The ions

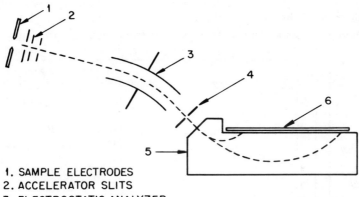

1. SAMPLE ELECTRODES
2. ACCELERATOR SLITS
3. ELECTROSTATIC ANALYZER
4. BEAM MONITOR
5. MAGNETIC ANALYZER
6. PHOTO PLATE

Figure 1.   Schematic diagram of a double focusing spark source mass spectrometer

produced are accelerated by a constant potential of 25 kV through the source slits into an electrostatic radial field which functions as an energy filter. As the ions pass through the magnetic field, the ion beam is split by deflection according to the mass-to-charge ratio. These charged particles impinge in focus on an ion-detector (photographic plate or electron multiplier) to form the mass spectrum. Elemental identification and abundance measurements can be made from the position and the relative intensity of the lines when the total ion beam current is known. This total ion current is measured by a monitor located just ahead of the magnetic analyzer. The resolution of the AEI-702R instrument used was

greater than 3000. Ilford Q-2 photographic emulsions were used to record
the mass spectra.

Conducting electrodes for general scan analyses were prepared by
mixing the pulverized coal or fly ash with an equal amount of pure silver
powder (99.999% silver). The homogenized mixture was then pressed
into polyethylene slugs in an isostatic electrode die at 25,000 psi for 1 min.
The nominal electrode size was 1 × 0.15 cm. Sets of graded exposures
were made from these electrodes so that the concentration range from
0.03 ppm up to the percent range was covered. Photoplates were interpreted according to the techniques given by Kennicott (1). In fly ash
and coal samples, iron was determined chemically so the iron isotopes
could be used as an internal standard. Computerized sensitivity values
were used (2).

Isotope Dilution By Spark Source Mass Spectrometry. A unique and
quite different approach to determining trace elements in solids, liquids,
and gases uses the isotope dilution technique. This method has been
operational for some time with mass spectrometers. Thermal ionization

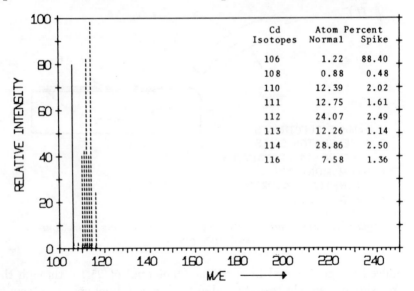

Cd Isotopes	Atom Percent Normal	Spike
106	1.22	88.40
108	0.88	0.48
110	12.39	2.02
111	12.75	1.61
112	24.07	2.49
113	12.26	1.14
114	28.86	2.50
116	7.58	1.36

Figure 2.   Computer plot of a cadmium spectrum spiked with enriched
^{106}Cd

sources are used for solids and electron bombardment sources for gases,
giving accurate results with small samples. However, it has not been
used until recently for analyzing environmental samples with spark
source mass spectrometers (3). The general method of isotope dilution is
described by Hintenberger (4). For each element to be determined, an
enriched isotope, usually of minor abundance, is mixed with the sample.

The isotopic ratios, altered by spike additions, are then measured on a portion of the sample by mass spectrometry. Even though the method is limited to elements having two or more naturally occurring or long lived isotopes, it is very sensitive and accurate and relatively free from interference effects. Thus, it has a great advantage over other analytical techniques. Thermal source and electron bombardment mass spectrom-

Table I.   Sample Computer Output for Cadmium by Isotope Dilution Spark Source Mass Spectrometry

CD	106	114	Vol	Conck
Spike	88.400	2.500	1.000	1.000
Sample	1.220	28.900	1.000	

SAMPLE	Nano-gm	106	114
5 Soil 318R	4005.000	19.3	16.2
5 Soil 318R	4032.000	46.9	41.3
5 Soil 318R	3933.000	64.6	60.0

eters are well suited for isotope dilution, but they have very different sensitivities for various elements. Spark source mass spectrometers have similar sensitivities for all elements and therefore can be used without deleterious effects from a complete matrix change.

In order to obtain, for example, the cadmium concentration in a fuel by isotope dilution SSMS, cadmium enriched in ^{106}Cd is equilibrated with the cadmium in the sample. The success of the technique depends on establishing isotopic equilibrium between the highly enriched ^{106}Cd and the normal cadmium in the sample. Isotopic and chemical equilibrium is attained by an acid (perchloric acid–nitric acid) reflux digestion and oxidation of the organic matter. Thereafter, any technique that permits the transfer of 0.1–3 ng of cadmium from the solution to the surface of a suitable substrate may be used. In this IDSSMS work, graphite was the substrate for cadmium, lead, and zinc, and copper was used for mercury (5).

For low metal concentrations, any suitable extractant may be used to concentrate the metal after equilibration with the enriched spike. High extraction efficiencies are not required since at this point the analysis depends on establishing a ratio between the enriched spike isotope and one of the major isotopes of the metal being sought.

Figure 2 shows a mass spectrum of cadmium spiked with enriched ^{106}Cd. The solid line at position 106 represents the ^{106}Cd spike, and the dashed lines represent the relative abundance for the other cadmium isotopes. The dashed line at juxtaposition at 106 is the relative abundance of ^{106}Cd as it occurs in nature. Table I shows the IBM 1130 computer-programmed output for a typical isotope dilution analysis. The program

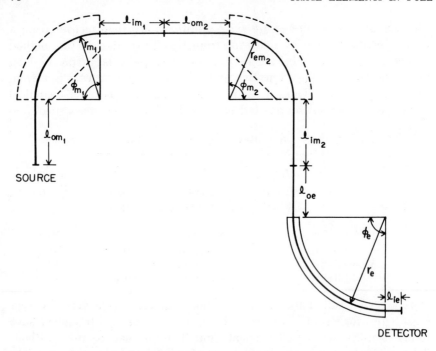

Figure 3.   Schematic diagram of the three-stage mass spectrometer. Parameters are designated as follows: r_{m_1}—30.48 cm; r_{m_2}—30.48 cm; r_e—43.26 cm; ϕ_{m_1}—90.00°; ϕ_{m_2}—90.00°; ϕ_e—90.00°; l_{om_1}—30.48 cm; l_{om_2}—30.48 cm; l_{oe}—24.28 cm; l_{im_1}—30.48 cm; l_{im_2}—30.48 cm; l_{ie}—7.77 cm.

is flexible in that the spike size, sample size, spike composition, and percent transmittance can vary. The three sets of data under 106 and 114 are percent transmittance for three photoplate exposures taken, and the results are reported in ng/g.

Isotope Dilution By Thermal Emission Mass Spectrometry. A three-stage thermal emission mass spectrometer (TEMS) was used for quantitatively measuring lead and uranium in coal and fly ash and lead in gasoline (Figure 3). The basic design of the instrument is modeled on that developed by White and Collins, 1954 (6) and modified at ORNL. The addition of an electrostatic third stage increased the abundance sensitivity to 10^8 as described by Smith *et al.* (7).

The two magnetic stages are a 30-cm radius followed by an electrostatic analyzer with a radius of 43.26 cm. The vacuum system is metal and is bakeable to 300°C. The analyzer region of the instrument is pumped with ion pumps and is maintained at a pressure 10^{-9} torr. A combination titanium sublimation–ion pump is used to obtain operating pressures in the source region in the 10^{-8} torr range. A Nier thick-lens source (8), used in conjunction with a sample wheel arrangement (9), makes it possible to analyze as many as 10 samples per day.

Ions are detected by a secondary electron multiplier behind the receiver slit. The pulses from the multiplier are accumulated in a 400-channel analyzer used in the time base mode. This arrangement allows the maximum sensitivity with respect to sample size. Mass measurements are made by sweeping the accelerating voltage across the region of interest—233–238 for uranium and 204–208 for lead. The determinations of uranium and lead in coal and fly ash and lead in gasoline, both quantitatively and isotopically, are done by isotope dilution with enriched spikes of ^{233}U and ^{204}Pb.

LEAD ANALYSIS BY TEMS. Lead analyses are made using the gel technique described by Cameron *et al.* (*10*). Lead ions are thermally produced at rhenium filament temperatures between 1100–1300°C, depending on the sample size. Ten ng of sample allows an analysis of 10 runs with 200 sweeps each across the masses of interest.

URANIUM ANALYSIS BY TEMS. Uranium ions are thermally produced from rhenium at filament temperatures 1700–1850°C. Using pulse counting for ion detection, small quantities of uranium (10–100 ng) in the form of uranium nitrate are loaded onto rhenium V–type canoe filaments, which produce enough ions for an analysis. After loading the filaments into the instrument, the source region is pumped down to a pressure of 1×10^{-5} torr, and the filaments are carburized *in situ* with benzene vapor, which is introduced through a Granville–Phillips variable leak. The resulting carburized filaments reduce the UO^+ and UO_2^+ ion signals to near zero and enhance the U^+ signal. This, coupled with pulse counting, greatly increases the measurement sensitivity. The precision of an isotope dilution analysis using this technique and where reagent blank contributions are kept insignificant is usually within ±1%.

PREPARATION OF COAL AND FLY ASH FOR ISOTOPE DILUTION ANALYSIS. Separate aliquots of coal and fly ash are weighed out and spiked with ^{204}Pb and ^{233}U, respectively. The chemical treatment and extraction of lead and uranium from coal and fly ash are identical, except coal is ashed at 450°C before chemical treatment. The samples are dissolved with a mixture of hydrofluoric, nitric, and perchloric acids in Teflon beakers. The lead is separated by dithizone extraction, evaporated to dryness, redissolved in dilute nitric acid, and 10 ng are loaded on filaments with silica gel for mass analysis.

The uranium is separated, after dissolving the sample as described for lead, by extraction with tributyl phosphate (TBP) from 4M nitric acid. After the organic phase is scrubbed with 4M nitric acid, the uranium is back-extracted into distilled water and evaporated to dryness. The uranium is loaded on a rhenium filament for analysis by dissolving the purified sample in a small volume of 0.05M nitric acid.

PREPARATION OF GASOLINE FOR ISOTOPE DILUTION ANALYSIS. Aliquots of gasoline are spiked with [204]Pb and treated by either a wet chemical method or the bromine oxidation method described by Griffing and Rozek (11).

The wet chemical procedure consisted of refluxing first with nitric acid until the initial reaction subsided. Then hydrofluoric acid was added to the flask, and the reflux continued until the reaction was complete. The resulting solution containing the inorganic lead was evaporated to near dryness, diluted to a suitable lead concentration with dilute nitric acid, and loaded onto rhenium filaments for M.S.

In the bromine method, bromine in carbon tetrachloride was added to the gasolines in a test tube. Heating assured complete conversion to lead bromide. The resulting lead bromide precipitate was dissolved with dilute ~1M nitric acid. The mixture was centrifuged, the organic layer discarded, and the aqueous solution was adjusted for M.S. analyses. Either of the procedures is satisfactory, but the bromine method is much easier, faster, and has less possibility for contamination.

Test Results and Discussion

The bulk of the samples for this study came from TVA's Allen Steam Plant at Memphis, Tenn. The sampling points (Figure 4) included inlet air, coal, bottom ash, precipitator inlet, and outlet at the 268-ft stack level. During the 2-week sampling period the unit was operated under steady state conditions at 240 MW (12) with a uniform coal supply so that a mass balance might be established for a number of elements. All the coal from southern Illinois was washed and crushed so that 90% was less than 4 mesh. Nominal coal analysis indicated the following composition: 9.5% moisture, 34% volatiles, 43% fixed carbon, 13% ash, and 3.4% sulfur.

The isotope dilution results in Table II are on fuel source samples obtained from NBS which were considered homogeneous. The results in Table III are from the sampling points indicated in Figure 4. These summarized results are mostly by the SSMS general scan technique which has an estimated accuracy of better than ±50%. The isotope dilution measurements are limited by the emulsion detector to ±3–5%. The results are in grams of metal flow per minute. The mass balance for the various elements was computed by the following equations:

$$Q_c(m) = C_c(m) \times (g \text{ coal/min}) \tag{1}$$

$$Q_{Pi}(m) = C_{Pi}(m) \times (g \text{ fly ash/min}) \tag{2}$$

$$Q_{ba}(m) = C_{ba}(m) \times (g \text{ ash in coal} - g \text{ fly ash/min}) \tag{3}$$

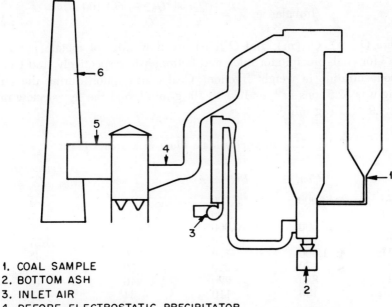

1. COAL SAMPLE
2. BOTTOM ASH
3. INLET AIR
4. BEFORE ELECTROSTATIC PRECIPITATOR
5. AFTER ELECTROSTATIC PRECIPITATOR
6. STACK SAMPLE

Figure 4. Schematic diagram of a TVA coal-fired steam plant

Table II. Isotope Dilution Results

Sample Type	*U TEMS*	*Pb TEMS*	*Pb SSMS*	*Cd SSMS*
		Concentration (wt ppm)		
Coal	1.22		28	0.28
	1.18		26	0.34
	1.22		28	0.32
	1.22		30	0.32
Fly ash	11.9		79	1.8
	11.7		78	2.2
	11.6		79	1.9
	11.6		76	1.5
Gasoline[a]		285[b]	284[c]	<.01
		282	284	<.01
		284	285	—
		283	279	—
Fuel oil[a]			0.27	0.002
			0.28	0.021
			0.26	0.018
			0.23	—

[a] Results in μg/ml.
[b] Wet oxidation.
[c] Br_2 oxidation with TEMS.

$$\text{Imbalance, \% } = \frac{(Q_{Pi} + Q_{ba} - Q_c) \, 100}{Q_c} \tag{4}$$

where $Q_c(m)$, $Q_{Pi}(m)$, and $Q_{ba}(m)$ are flow rates of metal (m) in g min^{-1} for coal, precipitator inlet, and bottom ash, respectively, and $C(m)$ is concentration in weight fraction. Coal consumption during the sampling was 82.5 tons hr^{-1}, or 1.25×10^6 g min^{-1}, and the fly ash flow rate was 4.9×10^4 g min^{-1}.

Table III. Impurity Mass Flow in a Steam Plant[a]

Element	Coal	Bottom Ash	Precipitator Inlet	Imbalance %
Al	13,000	5,500	7,300	−1.5
Ca	6,000	3,300	1,500	−20
Fe	25,000	10,000	4,900	−40
K	700	550	340	+27
Mg	1,800	770	340	−38
Na	370	220	150	0.0
Ti	880	220	240	−48
Mn	130	110	34	+11
As	6.2	0.22	2	−64
Be	<5	<1.1	0.83	—
Cd[b]	0.63	0.30	0.28	−7.9
Cu	63	22	19	−35
Pb[b]	9.3	0.45	10	+12
Ni	<100	55	24	−21
Sb	—	0.8	0.5	—
Se	7.5	2.2	1	−58
V	37	11	17	−24
Zn[b]	110	2.0	100	−7.3
Hg	0.08[c]	0.007[c]	0.007[b]	[d]

[a] Flow rate in g min^{-1}
[b] Isotope dilution
[c] Atomic absorption
[d] Most Hg flow is in stack gas

The metal balance for all the elements analyzed by mass spectrometry is good, but on the average shows a negative imbalance of 20%. Metals showing high imbalance, i.e., mercury, arsenic, and selenium, probably were in the gaseous state at the sample points. For example, a precipitator outlet sample of mercury showed a flow of 0.02 g min^{-1}. At this sampling point the particulates are much cooler than the fly ash at the precipitator inlet.

The average imbalance for the elements measured by isotope dilution mass spectrometry with lower volatility showed an imbalance range of −8 to +12%. These results indicate the usefulness of mass spectrometry in evaluating environmental impacts.

Acknowledgments

The authors acknowledge the assistance and cooperation of the many personnel of the Tennessee Valley Authority. This work was supported by the National Science Foundation RANN Environmental Aspects of Trace Contaminants Program under NSF Interagency Agreement AG–398 with the U.S. Atomic Energy Commission. The Oak Ridge National Laboratory is operated by the Union Carbide Corp. for the U.S. Atomic Energy Commission.

Literature Cited

1. Kennicott, P. R., "Interpretation of Mass Spectrograph Plates," in "Trace Analysis by Mass Spectrometry," (A. J. Ahearn, Ed.), pp. 179–210, Academic, New York, 1972.
2. Carter, J. A., Sites, J. R., unpublished relative sensitivity data, 1972.
3. Carter, J. A., Sites, J. R., "Environmental Spectrometry," *Annu. Chem. Div. Progr. Rep.* (Sept. 30, 1972) ORNL **4838**, 44–46.
4. Hintenberger, H., "A Survey of the Use of Stable Isotopes in Dilution Analyses," *Electromagn. Enriched Isotop. Mass Spectrom.* (1956), 177–189.
5. Carter, J. A., Sites, J. R., *Anal. Lett.* (1971) **4**(6), 351.
6. White, F. A., Collins, T. L., *Appl. Spectros.* (1954) **8**, 169.
7. Smith, David H., Christie, W. H., McKown, H. S., Walker, R. L., Hertel, G. R., *Int. J. Mass Spectrom. Ion Phys.* (1972/1973) **10**, 343.
8. Nier, A. O., *Rev. Sci. Instrum.* (1947) **18**, 398.
9. Christie, W. H., Cameron, A. E., *Rev. Sci. Instrum.* (1966) **37**, 336.
10. Cameron, A. E., Smith, D. H., Walker, R. L., *Anal. Chem.* (1969) **41**, 525.
11. Griffing, Margaret E., Rozek, Adele, Snyder, L. J., Henderson, S. R., *Anal. Chem.* (Feb. 1957) **29**, 190.
12. Bolton, N. E., Van Hook, R. I., Fulkerson, W., Carter, J. A., Lyon, W. S., Andren, A. E., Emery, J. F., "Trace Element Measurements at the Coal-Fired Allen Steam Plant Progress Report," **ORNL-NSF-EP-43**, March 1973.

RECEIVED January 21, 1974

8

Trace Elements in Coal by Neutron Activation Analysis with Radiochemical Separations

J. KENNEDY FROST, P. M. SANTOLIQUIDO, L. R. CAMP, and R. R. RUCH

Illinois State Geological Survey, Urbana, Ill. 61801

Procedures for the determination of 11 elements in coal—Sb, As, Br, Cd, Cs, Ga, Hg, Rb, Se, U, and Zn—by neutron activation analysis with radiochemical separation are summarized. Separation techniques include direct combustion, distillation, precipitation, ion exchange, and solvent extraction. The evaluation of the radiochemical neutron activation analysis for the determination of mercury in coal used by the Bureau of Mines in its mercury round-robin program is discussed. Neutron activation analysis has played an important role in recent programs to evaluate and test analysis methods and to develop standards for trace elements in coal carried out by the National Bureau of Standards and the Environmental Protection Agency.

Combustion of coal and other fossil fuels is a major source in the environment of trace elements that are hazards to human health. Toxic elements such as Hg, As, Sb, F, Se, and Tl are volatilized during coal combustion and are emitted directly into the atmosphere or concentrated in the fly ash (*1, 2, 3*). Most elements in coal occur at only parts per million levels, but large tonnages of coal are consumed each year in the United States. In addition, coal conversion processes, which could vastly increase coal use are now being considered seriously. The fate of trace elements during these processes is largely unknown.

There is, therefore, much interest in determining the concentration of various trace elements in coal. The Illinois State Geological Survey recently concluded a study of the occurrence and distribution of potentially volatile trace elements in coal sponsored by the U. S. Environmental

Protection Agency (EPA) (4). Several techniques were used to analyze 101 coals, mostly from Illinois, for 23 trace elements. Neutron activation analysis (NAA) was used to determine several of the elements.

The γ-ray spectrum of a coal sample irradiated with thermal neutrons is dominated initially by the short-lived isotopes, ^{38}Cl, ^{56}Mn, ^{24}Na, and ^{28}Al, and subsequently by such longer-lived isotopes as ^{46}Sc, ^{59}Fe, and ^{60}Co. By selection of a suitable irradiation period, decay interval, and counting period after irradiation, chlorine (5), sodium (4, 5), and manganese (4) can be determined by instrumental NAA with a NaI(Tl) detector and a 400-channel, pulse-height analyzer. This is the counting equipment available at the Illinois State Geological Survey and is the minimum equipment in most NAA laboratories.

If a solid-state Ge(Li) detector and a 1000- to 4000-channel analyzer are available, instrumental NAA can be extended to many elements. For example, Rancitelli (6) has analyzed coal and fly ash for 25 major, minor, and trace elements by using instrumental NAA with computer data reduction. Block and Dams (7) used similar analysis and reported on 43 elements in coal. Clearly, this is a good method for rapidly monitoring the composition of many coal samples.

Radiochemical separations are necessary for many elements when only a NaI detector is available. Even with a Ge(Li) detector, a radiochemical separation increases the sensitivity and accuracy and permits the determination of some elements whose radioactivities are masked by stronger activities in the multi-element spectrum of a coal sample. For example, mercury, selenium, gallium, and zinc in most coals are below the limit of detection instrumentally even with the resolution of a Ge(Li) crystal (7), but can be determined after radiochemical separations as is described later.

The increase in accuracy afforded by a radiochemical separation is absolutely necessary in the determination by NAA of trace elements in the coals selected as standards. The fact that interferences from the coal matrix are removed by a radiochemical separation is the advantage of this method of analysis over such instrumental methods as x-ray fluorescence and emission spectroscopy.

This article presents a comprehensive view of the present state-of-the-art of radiochemical separations for the following trace elements in coal: Hg, Rb, Cs, Se, Ga, As, Sb, Br, Zn, Cd, and U. Most of the work on the determination of trace elements in coal is very recent. The accuracy of the analysis methods, nearly all newly developed, has been open to question because of the lack of standards and lack of knowledge of the range of concentrations for many trace elements in coal. Federal government laboratories have taken the lead in evaluating methods of analysis and in developing standards. By a round-robin program, the Bureau of Mines evaluated analysis methods, including NAA, for determining mercury in coal. More recently, *via* another round-robin program, the National Bureau of Standards (NBS) and EPA evaluated methods of analysis and accuracy of results for several trace elements in coal. NBS is developing reference coal samples and testing methods of analysis, including NAA, to determine trace elements in coal.

Development and Evaluation of Radiochemical NAA by National Laboratories

Bureau of Mines Round-Robin for Mercury Determination in Coal. Mercury was the first trace element in coal to arouse environmental concern, prompted by data such as those of Joensuu (8). He reported that as much as 33 ppm mercury occurred in coal and inferred that coal combustion might be a major source of mercury in the environment. Lower values were reported by Ruch *et al.* (9) in the same year. Sixty-six coals analyzed by them contained 0.02–1.2 ppm mercury, and the mean mercury concentration of 55 Illinois coals in the set was 0.18 ppm.

In an effort to outline the problems arising in mercury determinations and to evaluate the methods used, the Bureau of Mines in 1971 organized a round-robin program involving 11 coal samples distributed to 20 participating laboratories (10). NAA with radiochemical separation and four analytical approaches using atomic absorption for determining mercury in coal were eventually evaluated. All five methods gave accurate results, but a combustion–double gold amalgamation–atomic absorption method was considered the best one available because it appeared more accurate and precise than the others and was quite simple, fast, and inexpensive. Results from the 14 laboratories reporting were used to calculate probable or best values of the mercury contents of the 11 coals, ranging from 0.05 to 0.41 ppm mercury. Of 100 NAA results reported, only 3% were rejected as inaccurate. Schlesinger and Schultz commented that activation analysis was a recommended procedure for determining mercury in coal if the nuclear reactor and other necessary expensive facilities were available and if the relatively long interval between sampling and results and the associated expense were not serious disadvantages. Details of activation analysis procedures used were not outlined.

NBS Standards and Analysis Methods for Trace Elements in Coal. In November 1971, NBS issued Standard Reference Material (SRM) 1630, Mercury in Coal, with a provisionally certified mercury content of 0.13 ppm. Later the provisional value of 2.1 ppm selenium in SRM 1630 was issued.

In mid-1972, an extensive laboratory intercomparison program was initiated by NBS and EPA to determine the accuracy of the current methods of analysis for trace elements in fuels, with the intent of improving the reliability of such determinatons. About 50 laboratories, using a variety of methods, participated in the analysis of a sample of coal, fly ash, fuel oil, and gasoline for 18 elements, As, Cd, Cr, Cu, Hg, Mn, Ni, Pb, Se, Tl, Th, U, V, Zn, F, Be, S, and Fe. NAA with radiochemical separation proved to be an important technique for determining a few elements, including mercury, arsenic, selenium, and zinc. NBS, in conjunction with this program, intends to provide a new SRM for each of the four matrices coal, fly ash, fuel oil, and gasoline, certified for 15 elements (11).

For a trace element concentration to be certified by NBS, it must be determined by at least two independent methods, the results of which must agree within a small experimental error range of ±1% to ±10%, depending on the nature of the sample and the concentration level of the element. Such accuracy in determining some trace elements for certification of coal SRM is achieved most easily by NAA with radiochemical separation. Scientists at NBS have extensively tested a neutron activation method that involves a combustion separation procedure on coal as well as on several other matrices to be certified as standard reference materials. The procedures they have thus developed to determine mercury (12), selenium (13), and arsenic, zinc, and cadmium (14) are outlined in a following section on methods for determining specific elements in coal.

General Considerations for Determining Trace Elements in Coal

Volatility of the Elements and Sample Treatment. A prime consideration in developing an analysis method is the volatility of the element to be determined. Controlled combustion of the coal sample and collection of the volatile products is a good way to separate very volatile elements such as mercury and bromine. The few completely volatile elements are subsequently and easily purified.

If the element to be determined is not volatile, it is advantageous to remove the organic material first by dry-ashing the coal at about 500°C. If the element is volatilized above but not below 150°C, the coal can be ashed at this low temperature by a radio frequency, oxygen–plasma asher (4). A low-temperature asher is expensive, and low temperature ashing, in particular, is time-consuming, but dry-ashing is a safe way to destroy the organic material. Moreover, the resulting ashes are easily brought into solution, either by mixed-acid digestion or alkaline fusion followed by dissolution in acid. Since fusion is rapid, it is particularly attractive in NAA when short-lived isotopes are determined. A further advantage of starting the NAA with a coal ash sample is that the trace element concentrations in the ash are as much as 10 or more times greater than in the coal.

When the whole coal is to be analyzed, bomb combustion may be safely used to destroy organic matter. Wet-acid digestion of coal is not without safety hazards. Gorsuch (15) discusses the problems encountered in decomposing organic material by acid digestion and by other methods. The oxidizing medium must be chosen carefully because some elements are volatilized from some acid mixtures. A nitric–sulfuric acid mixture is most commonly used to wet-ash coal. Greater precautions are necessary when perchloric acid is one of the oxidants used, and it is usually added after most of the organic matter has been destroyed by another oxidizing acid, *e.g.*, nitric.

After the coal or coal ash sample has been brought into solution, radiochemical separations may be made by any of several techniques, such as distillation, precipitation, solvent-extraction, ion exchange, etc.

Common Features of NAA Procedures. In all of the procedures discussed in this article, irradiations are made in a high thermal neutron flux (10^{11} to 10^{13} neutrons cm^{-2} sec^{-1}) simultaneously with the samples and standard(s) sealed in polyethylene containers for a short irradiation or in silica containers for a long irradiation. The standard is a known amount, or solution of known concentration, of a pure compound of the element to be determined. The concentration of the element in the sample is determined by comparing its radioactivity with that of the standard, which is either subjected to the same radiochemical separation as the sample with an inactive matrix or diluted. The radioactivity is counted directly if the sample is measured in solution. The radiochemical yield of precipitated samples is determined directly by weighing and that of solutions of samples by aliquot re-irradiation.

Methods for Determining Specific Elements

Mercury. This is the trace element in coal most studied by radiochemical NAA. The mercury value for SRM 1630 established by NBS resulted from a specific neutron activation with combustion procedure that Rook *et al.* (12) developed for biological samples.

The irradiated coal sample is placed in a ceramic combustion boat with added mercuric oxide carrier and burned in a slow stream of oxygen in a simple apparatus consisting of a straight quartz combustion tube connected to a straight condenser surrounded by a trap of liquid nitrogen. The ash and the tube are then heated to about 800°C to drive all volatile material into the cold trap. The products are dissolved in nitric acid. The activity of a vial containing a 2N HNO$_3$ solution of the collected material is counted with a Ge(Li) detector, measurements being made at the 0.077 MeV gold x-ray activity arising from the decay of 65-hr ^{197}Hg.

The only interference is from ^{82}Br which is, of course, more serious if counting is done with a NaI detector. Bromide is removed by adding bromide carrier to the nitric acid solution of the products at 40°C and precipitating with silver nitrate. Mercuric bromide is soluble in warm dilute nitric acid and is quantitatively retained in solution.

Tracer studies showed that mercury was recovered quantitatively from biological samples. Good precision and accuracy were demonstrated for the method by analysis of flour standards and several NBS biological standards.

Ruch *et al.* (9) used a combustion procedure to separate the mercury in their 1971 survey of 66 coals for mercury.

The irradiated sample, diluted with Alundum in a porcelain boat containing mercuric nitrate carrier, is combusted very slowly in a slow oxygen stream in a 96% silica combustion tube. The volatile products are collected in two consecutive traps, both containing a solution of acetic acid–sodium acetate buffer, bromine, and mercuric nitrate hold-back carrier. The collection solutions in 2N HCl are loaded onto Dowex 2, and radioactive interferences are eluted with aliquots of water and 2N HCl. The resin, in a small vial, is counted for the 0.077 MeV photopeak from ^{197}Hg.

The overall mercury recovery in the process is 67 ± 15%. The precision of the method is about 20%, and the detection limit is about 0.01 ppm mercury for a 1-g sample of coal. Reliability of the method was determined by the accurate analysis of two coal samples used in the Bureau of Mines study of the problems involved in determining mercury in coal (9) and then by the agreement within experimental error of the results from the 11 Bureau of Mines round-robin coal samples and their probable mercury contents (4).

More recently mercury has been determined at the Illinois Geological Survey (16) by a modification of the method of Rook *et al.* (12).

The coal sample is burned as described above, and the volatile products are collected in the straight tube condenser cooled by solid carbon dioxide. The collected material is dissolved in nitric acid, and ^{82}Br is removed by silver bromide precipitation. The resulting sample solution is counted for the 0.077 MeV activity of ^{197}Hg.

Radiochemical yields are 80–90%. The average relative standard deviation for the method is ±15%. The value of 0.14 ppm mercury was obtained on SRM 1630. The mercury contents of more than 100 coal samples from the United States have been determined at the Illinois State Geological Survey. Values range from 0.01 to 1.73 ppm (9, 16).

Pillay *et al.* (17) determined mercury by radiochemical NAA in a variety of environmental samples, including coal.

The irradiated sample and the polyethylene container in which it was irradiated, along with mercury carrier, are wet-ashed with a mixture of nitric, sulfuric, and perchloric acids under good reflux conditions. The mercury is isolated from the digest by precipitation as mercuric sulfide and is purified by electrodeposition as elemental mercury on gold foil, which is then counted for ^{197}Hg and 24-hr ^{197m}Hg x- and γ-ray activity with a thin NaI detector.

Chemical yields were generally 75–90%. The accuracy of the procedure was determined by tracer studies. Errors are less than 15% at the 0.01 ppm level and less than 5% at the 2 ppm level of mercury in biological tissues. Precision ranges from less than 17% relative standard deviation at the 0.01 ppm level to less than 5% at the 5 ppm level. Eleven coal samples from Ohio and Pennsylvania analyzed by this method had mercury concentrations of 0.32–1.20 ppm, with an average value of about 0.5 ppm (18).

O'Gorman et al. (19) used radiochemical NAA as a referee method in evaluating the determination of mercury by atomic absorption techniques. A commercial testing laboratory did the neutron activation analyses.

Irradiated coal samples were digested with mercury carrier in an acid solution, and the mercury was distilled as mercuric chloride. The mercury was electroplated from the distillate, and the activity was counted by multichannel γ-ray spectrometry.

Only one NAA determination was made on each of the 10 coals studied by O'Gorman et al. No estimate of the method's precision or sensitivity is given. Combustion–double gold amalgamation–atomic absorption results agreed well with the neutron activation results, and the former method was therefore considered to be more reliable than a combustion–solution–atomic absorption method that gave lower results.

Weaver and von Lehmden (20), under sponsorship of the EPA, evaluated two instrumental NAA methods and one with radiochemical separation for determining mercury in coal.

In the radiochemical procedure the irradiated coal sample and mercuric oxide carrier are digested with sulfuric acid, followed by nitric acid. Water and potassium bisulfate are added to drive off any nitric acid remaining. The mercury is separated by a standard dithizone extraction, and the extract is counted for the 0.077 MeV photopeak of ^{197}Hg with the NaI detector.

The three methods were evaluated by analyzing the 11 coal samples from the Bureau of Mines round-robin program. The radiochemical method proved to be reliable, but Weaver and von Lehmden noted that it is time-consuming, requires a large amount of laboratory equipment and fume hood space, and may have recovery errors. Instrumental NAA using a large-volume (36 cc) Ge(Li) detector did not give accurate results because of matrix interferences on the ^{197}Hg peak at 0.077 MeV. The other instrumental method in which counting was done on the newly developed 10 mm Ge(Li) low energy photon detector which has good resolution in the low energy region of the γ-ray spectrum and a 400-channel analyzer, gave good, fast results.

Oak Ridge is also using NAA extensively in its study of trace elements in coal and their disposition in power plant combustion (2). NAA with radiochemical separation is one method used for determining mercury in coal and coal ash. Irradiated coal samples are wet-ashed with a mixture of $HClO_4$, HNO_3, H_2SO_4, CrO_4^{2-}, and monochloroacetic acid, and coal ash with a mixture of HCl, H_2O_2, HF, HNO_3, and $HClO_4$. Mercury is separated as the sulfide, and its gamma activity is measured. Method reliability was determined by analysis of the 11 Bureau of Mines round-robin coal samples. The concentrations of some other elements in coal and coal ash determined by other methods are being checked at Oak Ridge by radiochemical neutron activation as well (2). The samples are wet-ashed as described above, and each element is separated specifically.

Rubidium and Cesium. The earliest study of trace elements in coal by radiochemical neutron activation was by Smales and Salmon, who reported on rubidium and cesium in 12 coals from County Durham, England in 1955 (21).

To determine rubidium, the irradiated coal sample and added rubidium carrier are digested and the organic matter destroyed by heating with nitric and sulfuric acids, followed by nitric and perchloric acids. The rubidium is separated from the gross activity of the sample by precipitation with ferric hydroxide. The rubidium is then precipitated as the cobaltinitrite and finally as the chloroplatinate. The activity of the chloroplatinate caused by the 1.8 MeV β^- particle of 18.7-day ^{86}Rb is counted with a Geiger counter.

In the determination of cesium, the coal sample and cesium carrier are digested, and the perchlorate separation and ferric hydroxide scavenging precipitation are made as in the procedure for rubidium. Cesium is then separated from the remaining solution by precipitation of cesium bismuth iodide. The final separation is made by precipitation of cesium chloroplatinate, which is counted with a Geiger counter for the 0.66 MeV β^- decay of ^{134}Cs ($t_\frac{1}{2} = 2.1$ yr), or counted for the 0.13 MeV γ-ray associated with the isomeric transition of ^{134m}Cs ($t_\frac{1}{2} = 2.9$ hr).

The chemical yield of each element was usually about 70%. The rubidium content of the 12 County Durham coals was 6–30 ppm, and the cesium content in 10 of the coals was 0.8–3.7 ppm. Reproducibility of results was very good for cesium; the relative standard deviation of the mean result was usually less than 5%. Rubidium results were slightly less reproducible, the largest relative standard deviation being 22%.

Selenium. In 1969, Pillay et al. (22) determined the selenium content of 86 coals from various coal-producing areas of the United States by NAA combined with extractive selenium distillation.

The irradiated coal sample with selenium carrier added is wet-ashed with a nitric–perchloric acid mixture under good reflux. Detailed tracer experiments were run to show that trace levels of selenium are not lost from the sample during ashing and equilibration of the carrier with the sample. The selenium(VI) is reduced to selenium(IV) and distilled into $2M$ HCl, with two additions of hydrochloric and hydrobromic acids. Elemental selenium is precipitated, dissolved in nitric acid and hydrogen peroxide, and reprecipitated. The ^{75}Se activity of the precipitate was counted with a NaI crystal.

The selenium recovery is usually about 80%. The authors (22) had good precision in their results. Selenium concentrations in the 86 coals ranged from 0.5 to 11 ppm but usually were from 1 to 5 ppm; the median was 2.8 ppm.

A similar procedure has been used at the Illinois State Geological Survey to determine selenium, but low-temperature coal ash was analyzed because it was determined that the amount of selenium that might be lost from coal on ashing at 150°C (sometimes up to 4–5%) was within the accuracy and precision of the experimental method (4).

The irradiated coal ash samples are digested under reflux with a mixture of hydrochloric, nitric, and perchloric acids. The selenium is then separated by the hydrobromic acid distillation and collected in water at 0°C. Elemental selenium is precipitated from the distillate with sulfurous acid, and the activity of the precipitate is counted. The photopeak arising from the 0.121 and 0.136 MeV γ-rays of [75] Se is measured.

Radiochemical yields are quantitative. The relative standard deviation of a measurement is usually better than ±10%. Analysis of SRM 1630 gave 2.0 ± 0.13 ppm selenium, and results for the NBS–EPA round-robin coal and fly ash samples agreed within experimental error with the probable certified values of selenium in those samples. The selenium concentrations of 101 coals analyzed by the above method range from 0.45 to 7.7 ppm and have a median value of 1.9 ppm (16).

At NBS, the neutron activation with combustion separation method used for determining mercury in coal (12) was further investigated for determining selenium by Rook (13). The same procedure is used except that the sample is heated finally to 1000°C. Mercuric oxide is also used as carrier for the selenium because selenium oxides are difficult to dissolve in mineral acids. The mercuric selenide formed carries the selenium effectively, and, as it is soluble in nitric acid, the dissolution procedure developed for the mercury separation can be used so that mercury and selenium can be determined in the same sample.

Rook counted the activity of his product solutions with a Ge(Li) detector and 2048-channel analyzer, measuring the 0.265 MeV γ-ray peak of 120-day ^{75}Se, as [197m]Hg interferes with the more commonly used 0.136 MeV peak.

Tracer studies showed that selenium is recovered quantitatively in the separation. The method is fast, and its precision and accuracy are good. The accuracy of a mean result is reported to be better than ±10% relative at the 95% confidence level. The provisional value of 2.11 ± 0.09 ppm selenium in SRM 1630 was established by this method.

Gallium. Since investigation showed that gallium is not lost when coal is ashed in a low-temperature plasma asher not exceeding 150°C, Santoliquido and Ruch (4, 23) determined gallium in the low-temperature ash of coal.

The irradiated ash sample and added gallium and zinc carriers are fused with sodium hydroxide. The melt is taken up in water, the mixture is filtered, and zinc hydroxide, which carries the gallium, is then precipitated. The precipitate is dissolved in 8M HCl solution, and the gallium is extracted with isopropyl ether and back-extracted with water. The activity of the aqueous solution is counted by measurement of the 0.832 MeV γ-ray photopeak of 14-hr ^{72}Ga.

Radiochemical yields are within 46–74%. The average relative standard deviation of the method is ±8%. The accuracy of the method was checked by analysis of a U. S. Geological Survey rock standard. Gallium concentrations in 101 coals, calculated from the concentration in the coal ash and the percentage of low-temperature ash in the coal, range from 1.1 to 7.5 ppm, the median value being between 2.9 and 3.0 ppm (16).

Arsenic. At the Illinois Geological Survey, arsenic has also been determined in low-temperature coal ash, and the concentration is calculated to a whole coal basis because it was found that negligible amounts of arsenic are lost in the low-temperature ashing of coal (4).

The irradiated ash sample, with arsenic carrier added, is digested under reflux with hydrochloric, nitric, and perchloric acids. Arsenic(III) is then distilled from the mixture with hydrobromic acid and collected in water. Elemental arsenic is precipitated with sodium hypophosphite, and the activity of the precipitate is counted. The 0.559 MeV γ-ray photopeak of 26.5-hr ^{76}As is measured. Radiochemical yields are quantitative.

The maximum relative standard deviation of a measurement found was ±12%. Results by this method for the NBS–EPA round-robin coal and fly ash samples agreed within experimental error with the probable certified values for the two samples. The arsenic concentrations in 101 coals analyzed ranged from 0.52 to 93 ppm (16).

The distillation separation procedure was the principal method used for the arsenic determinations, but more recently Santoliquido (24) has developed a method for the carrier-free separation of arsenic from low-temperature coal ash involving retention on an inorganic exchanger column.

The irradiated coal ash sample is fused with sodium hydroxide. A 7M HNO$_3$ solution of the melt is passed through a small chromatographic column filled with acid aluminum oxide or hydrated manganese dioxide. The column is rinsed with $7M$ HNO_3. The ^{76}As activity on the column is then counted.

Analyses of six coals gave results in agreement with those obtained earlier by the distillation method. The precision of both inorganic exchanger methods is good, with average relative standard deviations of less than ±8%. Results by the method using hydrated manganese dioxide are slightly higher than those by the distillation and acid aluminum oxide exchanger methods, which may reflect the fact that more metals are retained by the hydrated manganese dioxide exchanger.

At NBS, the neutron activation method with combustion separation step applied to the determination of mercury and selenium in coal has been modified and extended to analysis for arsenic, zinc, and cadmium by Orvini et al. (14).

After combustion of the sample and carriers in an oxygen stream, reducing conditions are achieved by a flow of carbon monoxide over the sample ash. Arsenic, zinc, cadmium, and any remaining selenium and mercury are reduced to elemental form. When the sample is heated to 1150°C in a slow carbon monoxide stream in a quartz tube in a furnace, recovery of all five elements in the liquid nitrogen trap is complete in 30 min. The recovery trap is washed with nitric acid to dissolve all the metals, and the radioactivity of a nitric acid solution of the products is counted with a Ge(Li) detector.

The method was tested by tracer experiments on several matrices, including coal and crude oil. Recoveries were quantitative or nearly so (97.8%). The method gave good results for zinc, cadmium, mercury, selenium, and arsenic in NBS orchard leaves and bovine liver and is currently being used to determine these elements in new SRMs.

Antimony. In the determination of antimony in coal, the irradiated coal sample mixed with benzoic acid and antimony trioxide carrier is burned in a Parr bomb (4).

The bomb contents are digested with concentrated hydrochloric acid, and material still undissolved is then digested with potassium hydroxide and hydrogen peroxide. A crude separation is made by a sulfide precipitation from the combined digestion solutions. The sulfides are dissolved in aqua regia, the solution is evaporated, and antimony in the residue is reduced to antimony(III) with hydroxylamine hydrochloride. The sample, in ammonium thiocyanate–hydrochloric acid medium, is loaded onto a Dowex 2 column (SCN⁻ form). Arsenic and other impurities are eluted with aliquots of more dilute ammonium thiocyanate–hydrochloric acid solutions. Antimony is eluated with sulfuric acid and fixed in solution by addition of hydrochloric acid. The activity of the solution caused by the 0.56 MeV γ-ray of 2.8-day ^{122}Sb is counted.

Dissolution of the oxides and/or other compounds of antimony present after combustion of the coal sample proved difficult. Radiochemical yields are rather low, ranging from 30 to 55%. The average relative standard deviation of an analysis is ±20%. The antimony content of 101 coals analyzed was 0.1–8.9 ppm (*16*).

Bromine. A NAA method for bromine in coal was developed (*4*) in order to have an interference-free method.

The irradiated coal sample, mixed with Alundum in a porcelain boat containing bromide carrier, is burned in the same manner as in the method for mercury (*4*), and the combustion products are trapped in two sodium or potassium hydroxide solutions. The alkali solution, in a large counting vial, is counted directly for the 0.56 MeV γ-ray photopeak of 36-hr ^{82}Br.

Chemical yields are 49–77%. The average relative standard deviation is ±10%. Bromine concentrations in 23 coals analyzed ranged from 4 to 29 ppm (*4, 16*).

Zinc and Cadmium. A radiochemical method was developed (*4*) for separating of zinc and cadmium from low-temperature coal ash.

The irradiated coal ash sample with zinc and cadmium carriers is fused with sodium hydroxide. The melt in 2*N* HCl solution is loaded onto a Dowex 1 anion exchange column (Cl⁻ form). The column is rinsed with 2*N* HCl, and cadmium and zinc are simultaneously eluted with distilled water. The activity of the eluate caused by the 0.438 MeV γ-ray of 13.8-hr ^{69m}Zn is counted immediately, and after a one-week decay period cadmium is measured by the 0.530 MeV γ-ray activity of 54-hr ^{115}Cd.

Radiochemical yields are 80–95% for zinc and quantitative for cadmium. The average relative standard deviation was ±25% for zinc and better than ±10% for cadmium. The detection limit of the method is 50 ppm cadmium in the ash. Analysis of two Illinois coals with un-usually high cadmium content (17 and 21 ppm) gave results in good agreement with those obtained by atomic absorption and by anodic stripping voltammetry (*4*). The recent development and testing of a radiochemical method for the determination of zinc, cadmium, and arsenic in coal and fly ash, by Orvini *et al.* (*14*), has already been discussed in the section on arsenic.

Uranium. Perricos and Belkas (*25*) have determined uranium in six coals by neutron activation followed by separation of the uranium daughter neptunium-239 by carrier-free extraction chromatography. The coals, from mines in northern Greece, had very high uranium concentrations (0.012–0.037%). However, uranium at the few parts per million level found in most coals could no doubt be determined by a modification of this method.

The coal samples are ashed at 600°C, and the ash is dissolved by digestion with nitric and hydrofluoric acids. Traces of hydrofluoric acid are evaporated, and dilute nitric acid solutions of the ash samples are irradiated. The irradiated ash solution, with nitric acid added, is heated to dryness, and the residue is taken up in hydrochloric acid. The solution is treated with hydroxylamine hydrochloride to reduce the neptunium quantitatively to the extractable neptunium(IV). The sample in hydroxylamine hydrochloride–hydrochloric acid solution is passed through a column of thenoyltrifluoroacetone in xylene as stationary phase on borosilicate glass powder as support, and the column is rinsed with hydrochloric acid–hydroxylamine hydrochloride solution. The neptunium is eluted with $6M$ HCl, followed by ethyl alcohol and $6M$ HCl, and the activity of the solution is counted by measuring the 0.106 MeV photopeak of ^{239}Np.

The average chemical yield of this separation had already been shown to be 99.5%, so no yield correction is necessary. The method is fast and has an error of about ±5%. The accuracy of the method was checked by analysis of two rock standards.

Summary

Neutron activation with radiochemical separation affords a reliable method for analyzing 11 trace elements in coal. Burning the coal and trapping the products is a simple way to separate certain elements from the organic matrix. The concentration of some elements in coal may be determined by analysis of the low-temperature ash of the coal. A variety of radiochemical separation methods can be used. There is also the possibility of developing methods of analysis for several other elements in coal. For example, ion exchange separations made with the newly developed inorganic exchangers present a largely unexplored field.

Literature Cited

1. Billings, C. E., Matson, W. R., Science (1972) 176, 1232.
2. Bolton, N. E., van Hook, R. I., Fulkerson, W., Lyon, W. S., Andren, A. W., Carter, J. A., Emery, J. F., "Progress Report," June 1971–Jan. 1973, Contract W-7405-eng-26, Oak Ridge National Laboratory, March 1973.
3. Swanson, V. E., Appendix J-II, "Report of the Coal Resources Work Group, Southwest Energy Study," U.S. Geological Survey, Jan. 1972.
4. Ruch, R. R., Gluskoter, H. J., Shimp, N. F., Ill. State Geol. Surv. Environ. Geol. Note (1973) 61, 43 pp.
5. Gluskoter, H. J., Ruch, R. R., Fuel (1971) 50, 65.
6. Rancitelli, L. A., "Pacific Northwest Laboratory Annual Report for 1972 to U.S. Atomic Energy Commission, Division of Biomedical and Environmental Research," Vol. II, Pt. 2, pp. 56–57, April 1973.
7. Block, C., Dams, R., Anal. Chim. Acta (1974) 68, 11.
8. Joensuu, O. I., Science (1971) 172, 1027.
9. Ruch, R. R., Gluskoter, H. J., Kennedy, E. J., Ill. State Geol. Surv. Environ. Geol. Note (1971) 43, 15 pp.
10. Schlesinger, M. D., Schultz, H., Bur. Mines Rept. Invest. (1972) 7609, 11 pp.
11. von Lehmden, D. J., Jungers, R. J., Lee, R. E., Jr., Anal. Chem. (1974) 46, 239.
12. Rook, H. L., Gills, T. E., LaFleur, P. D., Anal. Chem. (1972) 44, 1114.
13. Rook, H. L., Anal. Chem. (1972) 44, 1276.
14. Orvini, E., Gills, T. E., LaFleur, P. D., Trans. Amer. Nucl. Soc. (1972) 15, 642.

15. Gorsuch, T. T., "The Destruction of Organic Matter," Pergamon, England, 1970.
16. Ruch, R. R., Gluskoter, H. J., Shimp, N. F., "Illinois State Geological Survey Final Report," Contract No. 68-02-0246 and Grant No. R-800059, Mar. 1974.
17. Pillay, K. K. S., Thomas, C. C., Jr., Sondel, J. A., Hyche, C. M., *Anal. Chem.* (1971) **43**, 1419.
18. Pillay, K. K. S., Thomas, C. C., Jr., Sondel, J. A., Hyche, C. M., *Environ. Res.* (1972) **5**, 172.
19. O'Gorman, J. V., Suhr, N. H., Walker, Jr., P. L., *Appl. Spectrosc.* (1972) **26**, 44.
20. Weaver, J. N., von Lehmden, D. J., *Amer. Chem. Soc., Div. Fuel Chem. Preprint* **16**(3), 16 (1972).
21. Smales, A. A., Salmon, L., *Analyst* (1955) **80**, 37.
22. Pillay, K. K. S., Thomas, C. C., Jr., Kaminski, J. W., *Nucl. Appl. Technol.* (1969) **7**, 478.
23. Santoliquido, P. M., Ruch, R. R., *Radiochem. Radioanal. Lett.* (1972) **12**, 71.
24. Santoliquido, P. M., *Radiochem. Radioanal. Lett.* (1973) **15**, 373.
25. Pericos, D. C., Belkas, E. P., *Talanta* (1969) **16**, 745.

RECEIVED April 24, 1974

9

Trace Elements by Instrumental Neutron Activation Analysis for Pollution Monitoring

DEAN W. SHEIBLEY

Lewis Research Center, National Aeronautics and Space Administration, Cleveland, Ohio 44135

Methods and technology were developed to analyze 1000 samples/yr of coal and other pollution-related samples. The complete trace element analysis of 20–24 samples/wk averaged 3–3.5 man-hours/sample. The computerized data reduction scheme could identify and report data on as many as 56 elements. In addition to coal, samples of fly ash, bottom ash, crude oil, fuel oil, residual oil, gasoline, jet fuel, kerosene, filtered air particulates, ore, stack scrubber water, clam tissue, crab shells, river sediment and water, and corn were analyzed. Precision of the method was ±25% based on all elements reported in coal and other sample matrices. Overall accuracy was estimated at 50%.

The combustion of fuels, particularly coal, is a major source of trace element particulates emitted into the atmosphere. In 1970 alone, over 0.5 billion tons of coal, over 100 billion gallons of motor fuel, and nearly 60 billion gallons of fuel oil were burned in the United States (*1*). Trace levels of elements that are present in these fuels represent a potentially large contribution to the environmental burden, even if only a portion is injected into air.

Methods and technology were developed and used at the NASA Plum Brook Reactor (PBR) to analyze trace elements in pollution-related samples by instrumental neutron activation analysis (INAA). This work is significant because it demonstrates that INAA is a useful analytic tool for monitoring trace elements in a variety of sample matrices related to environmental protection. In addition to coal, other samples analyzed for trace elements included fly ash, bottom ash, crude oil, fuel oil, residual oil, gasoline, jet fuel, kerosene, filtered air particulates, various ores, stack scrubber water, clam tissue, crab shells, river sediment and water, cement, limestone, and corn.

Four goals were established for the INAA program at PBR. These goals were to:

1. Develop the technology and methods to analyze a large number of samples per year encompassing a variety of sample matrices.

2. Determine and report on as many elements as possible

3. Establish and maintain a high degree of accuracy and precision

4. Perform the work with minimum manpower and equipment, basically as a part-time effort.

To achieve these goals, the analysis scheme which involved sample preparation, irradiation, sample counting, and data reduction was optimized to achieve maximum sample output with minimum manpower expended.

Procedure and System Description

Capability Development. The work of Dams *et al.* (2) and Zoller and Gordon (3) was the basis for building the INAA capability. The INAA procedure they used involved the following steps.

The sample aliquots were encapsulated in polyethylene vials for the irradiation period. Two aliquots of the sample plus two standards were irradiated in a pneumatic transfer irradiation system, one set (sample plus standards) for a long time period (12–24 hrs) and the other for a short time period (5 min). After irradiation the samples were immediately removed from the vials. The short-time sample was counted after decay intervals of 3 min, 30 min, and 24 hrs. The long-time sample was counted after a decay interval of 3 wks; sometimes the decay intervals were 7–10 days. The typical neutron flux was 10^{13} neutrons/cm^2/sec. Counting data were processed through computerized data reduction codes. Twenty to thirty elements were reported.

Several major differences existed between this scheme and one compatible with the PBR facilities. The PBR did not have an operating pneumatic transfer irradiation facility, nor was there a sophisticated γ-ray spectrum analysis and data reduction computer program available. Irradiation facilities at PBR were hydraulic. Aluminum capsules (rabbits) were used to contain and transfer samples to and from the core.

The use of aluminum rabbits meant that data on the short-lived elements obtained from the 5-min decay count would be lost. Significant personnel radiation exposures were obtained from handling the aluminum rabbit ($\sim$ 50 g) directly from the reactor core because of the 1780 keV gamma of 2.2-min ^{28}Al and a remote method used to open the rabbit took too much time (in excess of 20 min).

High density polyethylene proved to be an acceptable rabbit material. With a 2.5-mm thick wall, it adequately withstood the 1.1×10^6

newton/m² (160 lb/in.²) hydraulic pressure of the PBR irradiation transfer system. The material did not significantly degrade in the PBR core
for periods up to 1 hr. The impurity level of the polyethylene was low.
As a result, radiation levels resulting from activation of the impurities
was slight, and these rabibts could be opened manually when returned
from the reactor core.

We now had acceptable rabbits for the short-term (polyethylene)
and long-term (aluminum) irradiations. Since the internal volume of
this rabbit design was cooled by primary cooling water flowing through
the rabbit, samples were protected by encapsulation in polyethylene and
quartz vials for the short- and long-term irradiations, respectively.

Decay time was critical to the determination of elements from the
5- and 30-min decay counts, so we decided to use the rabbit irradiation
facilities with the highest thermal neutron flux (10^{14} n/cm²/sec) to build
up the specific activity of short-lived isotopes. The higher flux also provided a greater sensitivity.

Evaluation of PBR Capability. To evaluate our overall capability
at this point, 10 particulate samples collected on Whatman-41 filter paper
were encapsulated, irradiated, and counted. Results on 16 elements
were manually calculated. The average time expended per sample was
10 hrs.

The entire process was then examined to identify those parts which
could be improved. This resulted in significant manpower savings and
more elements reported.

The analysis scheme for the 10 evaluation samples used two aliquots
($\sim$ 25 cm² of filter paper/aliquot). One aliquot was encapsulated in
polyethylene and irradiated in a polyethylene rabbit for 5 min in a
thermal neutron flux of approximately 10^{14} n/cm²/sec. This sample was
counted at decay times of 5 min, 30 min, and 24 hrs. The other aliquot
was encapsulated in high purity synthetic quartz and irradiated in an
aluminum rabbit 12–24 hrs. These samples were counted twice, after
decay periods of 10 days and 3 wks. Sample counting equipment included
one 4096-channel γ-ray spectrometer and a Ge(Li) detector.

The crystal of the Ge(Li) detector was 35 mm in diameter and 27
mm long. It had a nominal active volume of 20 cm³ and resolutions of
the 1332.5 keV photons of 2.18 (full width—half maximum) and 4.09 keV
(full width—0.1 maximum). Counting losses at 20% dead time were
about 6%.

The data in the memory of the γ-ray spectrometer was punched on
8-channel paper tape, converted to γ punched cards, and processed through
a rather primitive computer program which provided both a count per
channel output plus a not too reliable routine for peak finding and inte-
grating net area. All results were hand calculated from net peak areas and

theoretical nuclear data parameters. Standards were not used since they increased the number of items to be irradiated and counted. Sixteen elements were reported.

Manpower Analysis. Examination of the 10 man-hours/sample showed the following breakdown:

1. One-half hour was required to cut filter aliquots, load into vials, seal, and check the vials for leaks.

2. One hour was needed for the 5-min irradiation and counting after decay intervals of 5 and 30 min.

3. One-quarter hour was necessary for each count after decay intervals of 24 hrs, 10 days, and 3 wks.

4. The long-term irradiation operation procedure required only one-quarter hour for seven samples.

5. Eight hrs were required to hand calculate the data on 16 elements from computer-calculated net areas. The hand calculations were performed on programmable electronic calculators.

Evident shortcomings of the process were that the number of short-term irradiations per week was apparently limited by the available analyzer counting time and the manpower needed for data reduction. These two parts of the scheme were studied to improve efficiency by increasing the number of short-term irradiations and counting in an 8-hr shift as well as significantly reducing the man-hours for data reduction.

The solution to increasing the number of short-term irradiations was simple and also improved analyzer use efficiency. By irradiating two samples in the same rabbit, we could take advantage of the higher specific activity during longer decay times. The first sample was counted after 5 min decay and the second sample was counted immediately after the first at 10–12 min decay. The 30-min decay counts were performed after a decay time of 22–25 min on the first sample and 40–45 min on the second sample. Thus, short-term irradiation sample output was doubled while adding only approximately 0.25 hr to the original time interval of 1 hr/sample. All counting was performed on the same analyzer.

Data Reduction

The data reduction problem required a reliable computer data reduction program compatible with an existing IBM model 67 computer facility. The literature identified various codes: INVEN by Dams and Robbins (4), GAMANAL by Gunnick and Niday (5, 6), and SPECTRA by Borchardt *et al.* (7). INVEN could not be obtained in complete form. GAMANAL was written for the CDC 6600 computer and would require an extensive rewrite for the IBM 360. SPECTRA, although not too sophisticated, was quite adequate for this work. The peak find, peak integration, and peak identification properties were used as written. We inserted

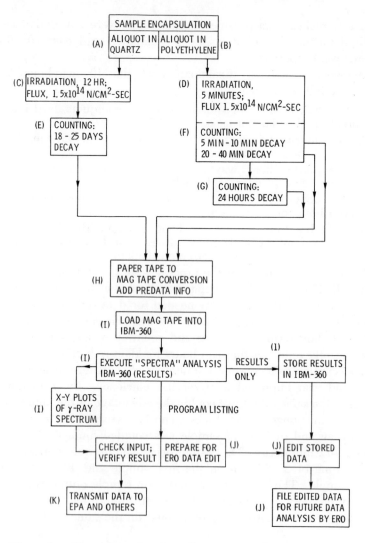

Figure 1. Flow chart showing scheme for irradiation, counting, and data reduction of various samples

the equations for hand calculations, since the code calculated results based on comparisons with standards. Our equations also permitted accurate corrections for dead time losses, decay during counting time, etc. A library of standards was added to the code based on elements found in the various sample matrices. The standard library eventually included 56 elements calculated from approximately 80 isotopes. The rewrite, program debugging, and irradiations of standards for the library took 7 months to complete.

The Optimum Capability at PBR. Twenty-two air particulate samples were analyzed to again evaluate the analysis scheme. The total time for sample analysis averaged 4.5 hrs/sample. Two hrs of the total were used to check computer input and results. Up to 56 elements were reported. At this point an initial goal had been achieved. We could analyze 12 samples/wk in approximately 54 hrs.

We continued our efforts for a more efficient operation. All isotope interference corrections, except for the aluminum correction on magnesium, were computerized. This saved more manpower in data reduction. By using routine scheduling of both long- and short-term irradiations and counting, we eventually achieved an optimum situation on both manpower and counting equipment availability. This optimum situation permitted analysis of 20–24 samples/wk with a total manpower of 3 to 3.5 hrs/sample expended.

Figure 1 shows the final scheme used for all sample types when the program was terminated. (The PBR ceased operation in Jan. 1973.) Each operation is identified by a letter in parentheses. Table I provides information on the manpower breakdown for each identified operation. Results on the pollution-related samples went to the Environmental Protection Agency (EPA), Division of Air Surveillance at Research Triangle Park, N. C. and the Environmental Research Office (ERO), NASA Lewis Research Center, Cleveland, Ohio.

A few comments should make Figure 1 more understandable. The sample aliquot used for the 5-min irradiation was sealed in a polyethylene vial, tested for leaks, and irradiated 5 min. The other aliquot was flame sealed in the quartz vial, tested for leaks, and irradiated 3–12 hrs, depending on the sample type. The counting data from each aliquot were

Table I. Optimum Manpower Breakdown for 20 Samples/Wk

Operation	Identification[a]	Average Manhours/Sample
Preparation, encapsulation	A	0.25
Preparation, encapsulation	B	0.25
Irradiation	C	0.1
Irradiation	D	0.1
Counting	E	0.25
Counting	F	0.55
Counting	G	0.25
Data handling	H	0.1
Data reduction	I	0.8
Data handling	{ J	{ 0.8
Data handling	{ or K	{ or 0.3
		Total 3.0–3.5

[a] See Figure 1.

punched on paper tape which was a permanent record of the counted aliquot. Input data, including the sample number and the requestor's identification number, were compiled. The input data and counting data were transferred onto magnetic tape after conversion to a format compatible with the IBM 360 computer. The output from the SPECTRA code consisted of a listing of all input data, counting data, and results, plus plots of the γ-ray spectrum of each sample count. EPA data listings contained a summary sheet of input data and results for data transmittal. Using a computer terminal the ERO results were stored in data sets in the IBM-360 memory, where they were edited for errors and refiled in memory for use in other ERO data reduction computer programs. The 56 elements reported are shown in Table II. They are grouped according to the decay time group in which they were determined.

Alternatives to Optimum Sample Processing. The 20–24 samples/ wk were considered the optimum because that number of samples could be completely analyzed in 5 days, working one $8\frac{1}{2}$-hr shift/day.

Another irradiation scheme used to reduce sample backlogs required performing 30–36 short-term irradiations during 1 wk, with the number of long-term irradiations per wk increased and held for counting 3 wks later. This approach involved longer range scheduling, was less routine, and was less efficient when unexpected reactor shutdowns occurred.

Problems Encountered

Some problems developed in sample preparation and irradiation because of the variety of sample matrices submitted for INAA. Some samples were particulates (coal, fly ash, bottom ash, ore), some were volatile hydrocarbons (gasoline, jet fuel, *etc.*), some were aqueous, and some were solids. Our methods of sample preparation were refined to provide all samples in sealed quartz and polyethylene vials for irradiation (*see* Figure 1). Depending on the contents of the quartz vials, the length of irradiation was varied from 3–12 hrs to minimize vial breakage from

Table II. Elements Listed in Data Summary According to Decay Group

≤15 min	>15 min ≤100 min		>100 min ≤5000 min			>5000 min			
Al	Ba	Mg	As	Gd	Sm	Ag	Hg	Se	Zn
Rh	Ca	Mn	Au	Ir	W	Ce	Lu	Sn	Zr
S	Cl	Na	Br	K		Co	Nd	Ta	
Ti	Dy	Sr	Cd	La		Cr	Ni	Tb	
V	Ge	Te	Cu	Mo		Cs	Rb	Th	
	I	U	Eu	Pt		Fe	Sb	Yb	
	In		Ga	Re		Hf	Sc		

Table III. Impurity Level of Whatman-41 Filter Paper

Element	μg/25.8 cm²	Element	μg/25.8 cm²
Ag	0.0088	Fe	1.57
Al	1.03	Hg	0.0046
Ba	0.049	Mn	0.018
Ca	3.5	Na	3.38
Ce	0.026	Sb	0.0046
Cl	3.23	Sc	0.0016
Co	0.0022	Ti	0.15
Cr	0.051	U	0.00067
Cu	0.08	Zn	0.601
Dy	0.00009		

pressure buildup caused by radiolysis of the samples. Minimum breakage occurred (less than 5 samples/100 irradiated) when volatile hydrocarbons (gasoline, kerosene, jet fuel) were irradiated for 3 hrs or less. Less volatile hydrocarbons and aqueous samples were irradiated 6 hrs. Solid samples were irradiated 12 hrs.

Another problem encountered was the impurity content of the filter paper used in the high volume samplers to collect the particulate samples. The conventional filter material used by EPA was glass fiber filter media. However, this was not compatible with INAA because of its high and varied impurity content. Discussions with K. Rahn of the Ford Reactor at the University of Michigan revealed that Whatman-41 filter paper was the most desirable medium for use with INAA (*see* Ref. 2). Our analyses showed Whatman-41 to be very low in impurities with consistent impurity levels from batch to batch. Average impurity levels, based on 12 batch analyses, are shown in Table III. Although the levels for calcium, chlorine, sodium, aluminum, and iron appear large, they rarely affected elemental levels found in filtered particulates. Impurity levels did not vary more than 25% from the mean.

Impurity levels were also important in the quartz and polyethylene vials. During the long irradiation time in quartz, the samples decomposed so much that it was impossible to completely remove the sample from the vial for counting. The samples had to be counted in the vials. We eventually determined that Suprasil (Suprasil T-20, 6-mm i.d., 8-mm o.d.; supplier, Amersil, Inc., Hillside, N.J.), a synthetic quartz, best served our needs for a low impurity vial material. In Table IV, weights of impurities are expressed in micrograms based on a vial 5 cm long, with a 6-mm i.d., 8-mm o.d., and weighing 2.77 g. For some elements the level varied significantly from one batch to another. The element which changed most significantly was antimony. The irradiation of volatile fuels for EPA required that we also determine the impurity level of polyethylene because liquid samples were counted in the irradiation vials. We found that impurity levels varied somewhat from batch to batch of polyethylene.

Table V presents average results on impurity levels of 16 poly-ethylene vials. Vials were 5 cm long, with 6-mm i.d., and 9-mm o.d. The average weight of the sealed vial was 2.00 g. High density polyethylene was used because of its better radiation resistance. The high impurity levels in these vials for S, Na, Cl, K, Al, and Ca severely limited the sensitivity to these elements in gasoline and other volatile materials which had to be counted in the vials.

Polyethylene was also susceptible to picking up radionuclides from the primary coolant. This interference was eliminated by sealing the sealed polyethylene vials into heat-shrinkable tubing which was easily removed after sample irradiation.

The problem of flame-sealing gasoline and other volatile materials into quartz for the long irradiation was solved by using a cold finger condenser in liquid nitrogen. The quartz tube (about one-half full, 0.8 cm^3) was positioned in an aluminum rod which had been bored out to accommodate the length of the tube from its bottom to above the liquid level. This rod was then lowered into a Dewar containing liquid nitrogen and allowed to stand until the gasoline became slushy and/or a ring of frost appeared just above the top of the aluminum. Then the tube was sealed using an oxygen–acetylene flame. This process took less than 5 min.

Table IV.   Typical Impurity Levels of a 2.77-g Suprasil Vial

Element	µg	Element	µg
Au	0.00035	Hg	0.00059
Ce	0.003	Sb	0.01
Cr	0.0144	Sc	0.0014
Co	0.0012	Ta	0.00017
Fe	0.52	Zn	0.092

Table V.   Impurity Levels in a 2.00-g Polyethylene Vial

Element	µg	Element	µg
Al	3.6	In	0.0003
Au	0.009	K	9.4
Ba	0.48	La	0.037
Br	0.25	Mn	0.046
Ca	4.5	Na	8.1
Cl	9.0	S	162.
Cu	0.27	Ti	0.95
I	0.062	V	0.016

Spectra Code

The computer program SPECTRA (7) contained an adequate peak find, peak integration, and peak matching routine. The code also distinguished two types of peaks based on the statistical significance of the γ-ray count data. The code resolved doublets and triplets using a minimum peak area of five channels (5 keV). This degree of resolution was adequate for almost all peaks used. SPECTRA also contained an option called G-search. The G-search routine was used to estimate the concentration of elements included as standards, but not found during the peak find and matching routine. The code then examined the energy region where γ-ray peaks should be located and provided three estimates of the element concentration.

These features were left intact. We did write into the code a library of standards which was used in lieu of irradiating and counting standards along with the samples. To establish our working standards, we used spectra of typical sample matrices. Element concentrations were calculated from nuclear data. We substantiated the elemental values of these matrix standards by comparison with NBS and other standards. Where discrepancies of greater than 20% existed, the data were examined, the problem identified, and corrective action taken. From this we concluded that for these sample types no significant matrix effects were present. We retained the original standards data tapes and occasionally reran them to check for error in the automatic data processing equipment and software.

Element standards were grouped into four standard libraries, corresponding to the four decay counting times. Decay time boundaries for each standard library are shown in Table II. In each library, at least two elements were calculated from different standards. These two standards represented different concentrations, counting geometries, dead times, decay times, and sample matrices. Visual inspection of the computer listing provided a rapid spot check for computer program malfunctions.

Another accuracy problem area involved the linearity of the γ-ray spectrometer (analyzer) detector system. SPECTRA control integers were adjusted to allow only a three-channel variation in γ-ray peak energy to ensure proper peak matching with standard peak energies. The linearity of the analyzer–detector system was checked daily with ^{137}Cs and ^{60}Co sources. The activity of these sources produced dead times in the analyzer–detector system of less than 10%. However, linearity drift did occur. Gain shifts in linearity also occurred when sample activities were too high. When the linearity shifted more than three channels, peaks did not match with standards or possibly could be improperly matched.

Table VI.   Interference Corrections for Various Isotopes

Element/ isotope (X)	Interfering isotope (Z)	Correction Required[a]
^{64}Cu	^{24}Na	NA(511 keV X) — 0.0886(NA 1369 keV Z)
^{203}Hg	^{75}Se	NA(279.1 keV X) — 0.3866(NA 264.4 keV Z)
^{65}Zn	^{46}Sc	NA(1115.4 keV X) — 0.7767(NA 889 keV Z)
^{141}Ce	^{59}Fe	NA(145.4 keV X) — 0.36(NA 1292 keV Z)
^{82}Br	^{99}Mo	NA(776.5 keV X) — 0.34(NA 740 keV Z)
Mg (g)[b]	Al (g)	g of Mg calculated — 0.057 (g of Al)
^{203}Hg [c]	^{175}Yb, ^{75}Se	See text

[a] NA refers to net area of peak.
[b] This correction was determined empirically but not programed into SPECTRA; all others were.
[c] Correction coefficients used were determined from a two-level factorial design.

A subroutine for linearity adjustment was added to SPECTRA to correct for nonlinearity or drifting of the analyzer system. The subroutine provided information on the true energy location of six peaks which could reasonably be expected in a given decay time group. The search area for each peak was restricted to a given number of channels. Coefficients for a linear equation were derived from a least-squares fit of the difference between true energy location and actual energy location of at least four peaks. The actual peak locations were then mathematically relocated closer to the true energy location before the peak matching routine with the standards was performed. This subroutine virtually eliminated missed peaks or improper peak matches. This saved considerable man-hours since peaks not matched with standards had to be manually identified and hand calculated.

Some elements were calculated from the same isotope in different decay groups. And other elements were calculated from different isotopes in the same or different decay groups. This duplication was used to improve accuracy.

Interference Corrections. In certain cases, the γ-rays of one element isotope are close to the γ-ray energy of an isotope used to determine another element. In these cases, it was necessary to correct for this interference. Subroutines for interference corrections were added to SPECTRA to eliminate performing such corrections manually. The correction factors used were normally derived from nuclear data information. But we did check these corrections by irradiating mixtures of these elements in various concentrations. The interference corrections used are shown in Table VI. The correction for ^{28}Al on ^{27}Mg was determined empirically because it was specific to the irradiation location in the PBR.

The determination of mercury in coal produced an interference correction problem which was quite complex. We found that counting an irradiation coal sample at 5–6-wks decay time produced a smaller mercury value than the value calculated from the count at 3-wks decay. A correction for ^{75}Se was being made on ^{203}Hg, but the lower mercury results (from 0.5 to 0.1 times smaller) at 6-wks decay could not be explained.

A search of the Nuclear Data Tables (8) produced another interference: 4.2-day ^{175}Yb. Not only did the 282-keV peak of ^{175}Yb interfere with ^{203}Hg, but also the 400.7-keV peak of ^{75}Se interfered with the 396-KeV peak area of ^{175}Yb, which was used for the Yb correction on ^{203}Hg. In addition, another ytterbium isotope, 32-day ^{169}Yb interfered with the 264-keV peak area of ^{75}Se, which is used in the correction on ^{203}Hg (peak area at 279 keV). These discrepancies were not eliminated by using theoretical corrections.

Finally, the problem was resolved by irradiating standards and mixtures of standards in a factorial experiment. The experiment design was a full factorial experiment with three variables, mercury, selenium, and ytterbium, at two levels with replication and with a center point added to test higher order effects. The pertinent information on treatments and levels of variables are shown in Table VII.

Regression analysis on the data was used to estimate the coefficients in a predictive model equation. The dependent variable was chosen as the difference between the computer-calculated value for mercury (or selenium or ytterbium) and the true value. Independent variables were the elements plus plausible interactions (*e.g.*, the interaction of selenium

Table VII.   Factorial Design Treatments for Yb and Se Interferences on Hg

| | *Coded Level of Variable*[a] | | |
Treatment	*Se*	*Hg*	*Yb*
1	−1	−1	−1
2	+1	−1	−1
3	−1	+1	−1
4	−1	−1	+1
5	+1	+1	−1
6	−1	+1	+1
7	+1	+1	+1
8	+1	−1	+1
Replicates			
4	−1	−1	+1
7	+1	+1	+1
8	+1	−1	+1
Center point	CP	CP	CP

[a] (−1) indicates low level, 10 μg. (+1) indicates high level, 100 μg. CP indicates $\left(\dfrac{\text{high level} + \text{low level}}{2}\right)$

Table VIII.   Comparison of NBS Standard

Element	NBS 610	PBR	NBS 612	PBR
Sb	—	—	—	—
Ce	—	—	(39)	37±2
Co	(390)	135±14	(35)	31±1
Eu	—	—	(36)	26±1
Au	(25)	20±2	(5)	5±1
La	—	—	(36)	35±15
Th	—	—	37.6±0.1	31±1
Sc	—	—	—	—
Ag	(254)	180±80	22.0±0.8	31±7

[a] NBS values in parentheses are interim values. Others are certified values.

and ytterbium on mercury). The coefficients derived for the predictive equations served as the basis for the empirical correction of ytterbium and selenium on mercury.

For comparison, the theoretical and empirical corrections are as follows:

1. Theoretical corrections for selenium and ytterbium interferences on mercury

$$A_c = A_u - 0.0369 \ X$$
$$B_c = B_u - 0.118 \ A_c$$
$$D_c = D_u - 0.959 \ B_c - 0.387 \ A_c$$

2. Empirical form of selenium and ytterbium correction factors for interference on mercury

Table IX.   Comparison of PBR Mean Values

Element	Coal	
	NBS[a]	PBR
As	5.9±0.4	5.9 ±0.5
Co	—	—
Cr	22 ±2	19.0 ±0.8
Cu	18 ±2	14.1 ±0.9
Hg	0.11	0.95±0.09
Mn	47 ±3	38 ±3
Ni	—	—
Rb	—	—
Se	2.8±0.2	3.8 ±0.5
Sr	—	—
Fe	—	—
Th	3	3.1 ±0.2
U	1.4±0.1	0.98±0.08
V	35 ±4	36 ±4
Zn	37 ±4	—

[a] Data taken from: "Characterization of Standard Reference Materials For the 0158(D), File Number 0158-4 (May 1, 1973).

Reference Materials with PBR Results (ppm) [a]

NBS 614	PBR	NBS 616	PBR
(1.06)	1.1 ±0.1	0.078±0.007	0.012±0.02
—	—	—	—
0.73 ±0.02	0.59±0.1	—	—
0.99 ±0.04	1.1 ±0.6	—	—
(0.5)	1.0 ±0.8	—	—
0.83 ±0.02	<2	—	—
0.746±0.007	0.58±0.15	0.025±0.004	0.018±0.002
0.59 ±0.04	0.68±0.23	0.026±0.012	0.020±0.004
0.46 ±0.02	0.57±0.07	—	—

$$A_c = A_u - 0.0433\ X$$
$$B_c = 0.443\ (B_u - 0.118\ A_c)$$
$$D_c = D_u - 1.65\ B_c - 0.387\ A_c$$

where

A_c = selenium 264.6 keV area corrected for ^{169}Yb
A_u = net area of selenium (264.6 keV), uncorrected
X = net area of ^{169}Yb at 177.2 keV
B_c = ytterbium 396.1 keV area corrected for selenium (264.6 keV)
B_u = net area of ytterbium (396.1 keV) uncorrected
D_c = mercury 279.1 keV area corrected for selenium and ytterbium
D_u = net area of mercury (279.1 keV) uncorrected

on Round-Robin Samples with NBS Probable Certified Values (ppm)

Fly Ash		Fuel Oil	
NBS [a]	PBR	NBS [a]	PBR
61 ±3	69 ±8	—	—
38	38 ±4	—	—
132 ±5	122 ±12	—	—
120	142 ±9	—	—
0.15	3.7±1.1	0.002	0.022±0.015
495 ±30	466 ±31	0.12	0.19 ±0.00
—	—	37±3	39.5 ±2.3
112	115 ±15	—	—
9.3±1.0	12.7±1.8	0.14	0.19 ±0.03
1380	869 ±33	—	—
—	—	12	12.4 ±1.6
24	25 ±2	—	—
11.6	8.4±0.6	—	—
214 ±8	230 ±11	315	266 ±18
210 ±20	700 ±220	0.2	0.48 ±0.12

Determination of Trace Elements in Fuels," Bimonthly Report *NBS–EPA–1AG–*

A special subroutine, incorporating this correction and the tests for when it should be used, was added to SPECTRA. We checked it by irradiating and analyzing mixtures with known amounts of ytterbium, selenium, and mercury. The true values and the values calculated by the computer agreed within ±15% for mercury. The method allowed the determination of mercury in the presence of ytterbium and selenium when their concentration ranged up to 10 times greater than mercury.

Samples which were counted in polyethylene and quartz vials required corrections for the impurity content of the vials. Standard libraries of vial and blank filter paper corrections were added to SPECTRA (see Tables II, III, and IV). We used indicators in the input data to each computer calculation to call out the proper correction library. The code used corrections for polyethylene vials, Suprasil vials, Whatman-41 filters (25.8 cm²), and combinations. The computer also did not print the value for an element in a sample if the microgram quantity was within two times the microgram value of the vial or filter paper. The value output was listed as less than the vial or filter paper value, corrected to proper units. With this restriction, some data were lost, but very small values which were the difference between two larger numbers were eliminated. For example, if a volatile sample plus vial gave a chlorine value of 9.4 μg, the chlorine value output by the computer for the sample would be less than 9.0 μg (referring to chlorine in Table V) rather than the difference of 0.4 μg. If the sample plus vial gave a chlorine value of 20 μg, the value output by the computer would be 11 μg.

Precision and Accuracy

As a measure of accuracy, we checked our method against NBS standards and mixtures of elements of known concentration and also participated in a round-robin analysis with NBS and EPA. We analyzed four NBS standards containing 60 elements in glass. Comparisons of results are given in Table VIII. With the exception of the cobalt result in the NBS 610 Standard, agreement is generally within ±25% of the NBS value.

In the round-robin analysis, a minimum of five samples each of coal, fly ash, gasoline, and fuel oil were analyzed. The NBS Probable Certified Value for certain elements are shown in Table IX along with PBR values. Since no data were reported on gasoline, there are no comparisons. The most inconsistent comparison was for mercury. Only four laboratories reported mercury by INAA in coal, three by INAA in fly ash, and two by INAA for fuel oil. Most other laboratory results reported were based on atomic absorption spectrometry. With one exception, all mercury values reported by INAA (a nondestructive method) were higher than

Table X.   Precision on NBS–EPA Round-Robin Coal Sample

Element	Mean (ppm)	±1σ (ppm)	Standard Deviation (%)	Counting Precision Range at 1σ (%)
Al	15,700	1,550	9	0.6–1
As	5.9	0.5	9	10–12
Au	0.146	0.048	33	10–40
Br	20	3	15	9–12
Ba	337	42	12	5–8
Ca	4,070	560	14	8–15
Ce	17.340	0.089	2	1–2
Cl	750	75	10	2
Co	5.48	0.15	3	1–13
Cr	19	0.8	4	3–5
Cs	2.55	0.06	2.3	8–10
Cu	14.1	0.9	6	3–5
Dy	0.85	0.06	7	2
Eu	0.312	0.037	12	0.2–0.3
Fe	7,517	119	2	1
Ga	5.4	0.8	14	11–15
Ge	70	5	7	35–50
Hf	0.92	0.05	6	5–10
Hg	0.95	0.09	10	25–33
I	2.78	0.38	14	12–30
In	0.04	0.01	25	20–35
Ir	2.48	0.27	11	5–12
K	3,500	360	10	3–4
La	11.3	3.3	30	6–12
Lu	0.416	0.017	4	5–8
Mg	980	250	26	12–33
Mn	38.0	2.6	7	0.5
Na	370	33	9	2–3
Nd	6.4	1.5	24	40–55
Rb	19	1.9	10	10–20
Sb	6.4	1.6	24	8–15
Se	3.8	0.51	13	25–33
Sm	1.3	0.19	15	2–5
Sn	125	20	16	10–15
Sr	93	9.2	10	8–11
Ta	0.360	0.028	8	15–20
Tb	0.03	0	0	0.5–7
Th	3.1	0.2	8	2–20
Ti	1,312	150	12	10–20
U	0.980	0.078	8	8–12
V	36	4	11	5–10
W	1.9	0.8	40	30–80
Yb	0.55	0.04	8	15–20

those reported by destructive methods of analysis. The higher INAA results indicate that either isotopic interferences during sample counting were not properly corrected for or mercury was lost from the samples during preparation for destructive analysis methods. Results on most other elements fall within a ±50% accuracy when compared to the NBS value as the true value.

To demonstrate the precision of the INAA method at PBR, data are presented on the same NBS–EPA round-robin samples. These data are typical of trace element results reported to EPA and ERO. Trace element results on other matrices may be found in Refs. *9, 10,* and *11.*

Five aliquots of coal were analyzed to determine the mean value and the standard deviation. The column *Counting Precision Range* refers to the counting error associated with the peak area of the isotope used in the calculation. This counting error is described in Ref. 7. Results are reported on 43 elements in Table X. The elements antimony and lanthanum exhibit a larger per cent standard deviation than the range of counting precision. These elements might not be homogeneous with respect to the coal matrix. Precision at 1σ ranges from 2% for cerium and iron to 40% for tungsten.

The data on fly ash reported in Table XI was based on five aliquots. Forty-three elements are also reported here. Precision ranges from less than 5% for aluminum, sodium, strontium, and vanadium to 50 and 55% for magnesium and ytterbium, respectively.

The results on gasoline and fuel oil are presented in Tables XII and XIII, respectively. The results on gasoline for six elements are based on six aliquots. These results show that sulfur content of gasoline can be determined by INAA. The fuel oil results for 16 elements are based on seven aliquots. The sulfur content reported, 2.04%, agreed well with the NBS true value of 2.035% for the round-robin fuel sample.

The average precision based on the data in Tables X–XIII is 14.4 ±11.1% for all four matrices. Hence, the overall precision of the method is placed at ±25%.

The accuracy of the method is estimated at ±50%. This estimate not only includes the uncertainties in counting methods but also the uncertainties associated with nuclear parameters.

Conclusions

Four goals were realized for the pollution-related INAA program at the NASA Plum Brook Reactor:

1. Develop technology and methods to analyze a large number of various matrices per yr; the capability was geared for 1000 samples/yr.

Table XI.   Precision on NBS–EPA Round-Robin Fly Ash Sample

Element	Mean (ppm)	±1σ (ppm)	Standard Deviation (%)	Counting Precision Range at 1σ (%)
Al	109,600	4,020	3.7	0.06–1
As	69.5	7.6	11	8–10
Ba	2,734	167	6	6–8
Br	12.1	1.5	12	35–50
Ca	41,000	3,600	8.8	7–16
Ce	129	10	7	2–10
Cl	185	44	24	10–40
Co	38.6	3.7	9	1–10
Cr	122	12	10	2–4
Cs	13.8	1.4	10	8–12
Cu	142	9	6.2	3–4
Dy	7.6	2.4	31	1–8
Eu	2.42	0.16	6.6	1–2
Fe	52,780	5,600	11	1
Ga	38.3	6.3	16	8–20
Ge	476	166	35	—
Hf	7.62	0.56	7.3	3–20
Hg	3.7	1.1	30	15–30
In	0.156	0.035	23	33–50
Ir	18.6	3.3	18	8–10
K	21,800	2,400	11	3–4
La	77	8	10	3–8
Lu	3.8	0.5	14	5–8
Mg	15,970	8,060	50	15–33
Mn	466	31	6.6	0.3
Na	2,658	129	4.9	1–2
Rb	115	15	13	8–14
Sb	12.08	0.86	7	10–15
Sc	27.5	2.4	9	0.5–4
Se	12.7	1.8	15	20–30
Sm	10.05	0.58	5.8	2–3
Sn	740	210	28	8–20
Sr	869	33	3.8	5–8
Ta	2.74	0.25	9	15–20
Tb	0.22	0.04	16	3–6
Th	25	2	8	2–3
Ti	8,900	752	8.5	10–12
U	8.40	0.56	6.7	8–15
V	230	10.6	4.6	5–7
W	12.7	1.1	8.8	30–50
Yb	6.2	3.4	55	3–25
Zn	700	220	31	10–30
Zr	640	140	22	25–35

Table XII.   Precision on NBS–EPA Round-Robin Gasoline Sample

Element	Mean (ppm)	±1σ (ppm)	Standard Deviation (%)	Counting Precision Range at 1σ (%)
Br	128	10	8	1
Ca	8	4	50	30
Cl	80	6	7	2
Cu	2.3	0.2	8	10–12
Dy	0.0066	0.0020	31	15–40
S	335	44	13	25–40

Table XIII.   Precision of NBS–EPA Round-Robin Oil Sample

Element	Mean (ppm)	1σ (ppm)	Standard Deviation (%)	Counting Precision Range at 1σ (%)
Au	0.0245	0.0007	3	15
Br	0.24	0.07	30	35–40
Ca	15	2	15	—
Cl	18	0.7	4	—
Co	0.25	0.01	4	1–20
Cr	0.116	0.035	30	8–30
Cu	0.22	0.02	10	—
Fe	12.4	1.6	13	5–12
Hg	0.022	0.015	66	10–33
Mn	0.19	0	—	8–12
Ni	39.5	2.26	6	3–7
S	20,400	3,900	19	15–20
Sb	0.0146	0.0033	23	16–33
Se	0.19	0.03	17	25–30
V	266	18	7	2–3
Zn	0.48	0.12	25	15–40

2. Determine and report as many elements as possible; up to 56 elements were routinely reported.

3. Establish and maintain a high degree of accuracy and precision using the INAA method; precision was normally less than ±25%, and accuracy was estimated at less than ±50%.

4. Perform the work with minimum manpower and equipment; 20–24 samples were analyzed/wk, 3–3.5 man-hours/sample, and only one analyzer–detector system.

The methods used to achieve these goals involved analyzing samples in large numbers; carefully planning and scheduling sample preparation, irradiation, and counting; complete computer automation of the data reduction scheme; and strict attention by the work team to small details of the overall scheme to avoid re-running samples because of equipment or method malfunction or data input errors.

The value of INAA for monitoring trace element pollutants resides in several demonstrated facts: the method adapts to a large variety of sample matrices with only small changes in techniques; it provides reasonably accurate data on a large number of elements; and work properly planned and executed need not absorb a large amount of highly skilled manpower.

Acknowledgments

The author wishes to acknowledge the efforts of the following NASA personnel for the success of this program. The sample preparation, irradiation, and counting were the responsibility of Robert Defayette, Clark Hahn, Sr., Ray Arndt, John Fryer, and Ronald Scott. Data on nuclear parameters, interference corrections, and standards were provided by Paul Richardson. Data checking, data editing, data changes, and interference correction validation were performed by Anne Bodnar. Computer programming and operation were performed by David Hopkins.

Literature Cited

1. Lee, R. E., Jr., Von Lehmden, D. J., *J. Air Pollut. Contr. Ass.* (1973) **23**, 853.
2. Dams, R., Robbins, J. A., Rahn, K. A., Winchester, J. W., *Anal. Chem.* (1970) **42**, 861.
3. Zoller, W. H., Gordon, G. E., *Anal. Chem.* (1970) **42**, 257.
4. Dams, R., Robbins, J. A., "Nondestructive Activation Analysis of Environmental Samples," COO-1705-6; Rept. **48**, University of Michigan, Ann Arbor.
5. Gunnick, R., Niday, J. B., "Computerized Quantitative Analysis by Gamma-Ray Spectrometry. Vol. I: Description of the GAMANAL Program," Rept. **UCRL-51061**, California University Radiation Laboratory, Livermore, March 1, 1972.
6. *Ibid.*, Vol. II: "Source Listing of the GAMANAL Program," Dec. 6, 1971.
7. Borchardt, G. A., Hoagland, G. E., Schmitt, R. A., *J. Radioanal. Chem.* (1970) **6**, 241.
8. "Nuclear Data Tables," Academic, New York.
9. Sheibley, D. W., "Trace Element Analysis of Coal by Neutron Activation," *Tech. Memo.* **TM X-68208**, National Aeronautics and Space Administration, Washington, D.C., August 1973.
10. Sheibley, D. W., "Trace-Element Analysis of 1000 Environmental Samples per Year Using Instrumental Neutron Activation Analysis," *Tech. Memo.* **TM X-71519**, National Aeronautics and Space Administration, Washington, D.C., March 1974.
11. Fordyce, J. S., Sheibley, D. W., "Estimate of Contribution of Jet Aircraft Operations to Trace Element Concentrations at or Near Airports," *Tech. Memo.* **TM X-3054**, National Aeronautics and Space Administration, Washington, D.C., May 1974.

RECEIVED January 21, 1974

10

Major, Minor, and Trace Element Composition of Coal and Fly Ash, as Determined by Instrumental Neutron Activation Analysis

K. H. ABEL and L. A. RANCITELLI

Battelle, Pacific Northwest Laboratories, Richland, Wash. 99352

A highly sensitive instrumental neutron activation analysis (INAA) technique can determine 30–40 elements in coal and fly ash including Sb, As, Se, Hg, Zn, and V which are elements of environmental concern. This multi-element capability makes possible the detailed chemical analysis of coal and combustion products. A comparison to National Bureau of Standards–Environmental Protection Agency fuel standards shows that a high degree of accuracy and precision is attainable, allowing tentative conclusions to be drawn as to the origin of certain elements in the coal and their fate during combustion.

The current emphasis on the large-scale exploitation of coal as a commercial electrical energy source makes a more thorough understanding of the chemistry of coal and the by-products of coal combustion increasingly important. An example of planned large-scale coal use for power generation is the proposed North Central Power Project. This project will use the vast coal reserves in the Fort Union Formation in North Dakota, South Dakota, Montana, and Wyoming. The coal resources in this area are estimated to be 40% of the U.S. total reserves. The North Central Power Study (*1*) estimates that by 1980 a steam-fired generating capacity of 53,000 MW will exist in an area between Colstrip, Montana, and Gillette, Wyoming. Each 1000 MW of power generated will require consumption of approximately 4 million tons of coal/yr. Thus, the total coal consumption will be approximately 210 million tons/yr. To help put this coal consumption in perspective, the Department of Interior estimated that 326 million tons of coal were used for electrical generation in the entire United States in 1970 (2). Therefore, the estimated 210 million tons represents approximately 65% of the total used for electrical generation in 1970.

To illustrate the possible impact of this type of development, some further calculations can be performed using present emission standards and representative coal concentrations. First, if one uses an average sulfur concentration of 0.65% typical of low sulfur western coals and assumes the estimated coal consumption, 2.7 million tons of sulfur dioxide will be produced per year. The emphasis in recent environmental impact studies of fossil fuel power plants has been on sulfur dioxide emission and associated acid rains which have had detrimental effects on surrounding vegetation (3, 4, 5). There are Federal standards to limit sulfur dioxide emissions, and the technology for sulfur removal required to meet these standards is developing in several alternative pathways, all still in experimental stages. Thus, there undoubtedly will be numerous modifications and alterations in plant design before sulfur dioxide is removed on a full-time and large-scale basis. But even under these standards, emission of 2.1 million tons of sulfur dioxide/yr would be allowed, assuming a coal consumption of 210 million tons.

Secondly, if it is assumed that the coal has an ash figure of 15% upon combustion (which appears reasonable for these coals) and 99% efficiency is assumed for the precipitators, 315,000 tons of fly ash could be released per year under the projected coal consumption.

Finally, the trace constituents become important factors when substantial amounts of coal are burned. Selenium, which is present at approximately 2 ppm in coal, could be released at a rate of 210 tons/yr, assuming 210 million tons of coal consumed and a 50% escape rate during combustion (6). Significant discharges could also be expected for Hg, F, As, and perhaps Sb, Zn, and Cu.

Thus it becomes apparent that, for the very high projected coal consumption in this region, the discharge of trace constituents is important, and potential environmental interaction and effects warrant immediate attention. There has been some attention given to radioactive elements in coal and their release during combustion (7, 8, 9). However, the stable element trace constituents are permanent pollutants of the environment, and their immediate and potential long-term effects should be investigated.

Research indicates that a significant fraction (50–90%) of mercury is volatilized and lost during coal combustion (10, 11, 12) and that many of the potentially hazardous trace elements appear concentrated upon finer particulate emissions (13, 14). Several investigators have observed enrichment of these hazardous elements upon particulates in urban areas

(*15–23*), in remote locations such as Antarctica (*24*) and Northern Canada (*25*), and in the upper atmosphere (*26*).

There are several factors which make neutron activation analysis (NAA) an appropriate technique for investigating potential pollutants in coal and the combustion process. First, the multi-element nature of NAA is useful because of the large number of potential elemental pollutants, such as Se, Hg, As, Zn, Ni, Sb, and Cd. Also, the use of elemental ratios made possible by the multi-element capability facilitates the understanding of chemical behavior during the combustion process. Elemental ratios have been used previously in urban (*15*) and upper atmospheric (*26*) studies. Secondly, the sensitivity and selectivity of NAA allows determination of many elements present at very low concentrations (ppm or lower), and the results are unaffected by matrix interferences. This sensitivity also allows analysis of very small samples. Finally, the cost of NAA when conducted as a multi-element analytical tool is competitive with more conventional and less sensitive techniques on the cost-per-element-per-sample basis.

Battelle has developed instrumental neutron activation analysis (INAA) techniques which permit very sensitive and accurate multi-element analysis of approximately 40 elements in coal and fly ash. These techniques, which will be described in this work, form the basis for extensive environmental studies of the effluent from coal-powered generating facilities and other pollution sources.

Experimental

These techniques are designed to minimize both the actual working time required and the analytical uncertainties in sample analysis. Sample preparation and neutron activation procedures are based on proved analytical and microanalytical techniques. The unusually high sensitivity, reliability, and accuracy are achieved through a choice of optimum irradiation and counting times for the γ-ray detection systems.

Sample Preparation. The coal and fly ash samples are weighed into small bags formed by folding and heat-sealing polycarbonate film. The openings are also heat-sealed after the samples are placed into the polycarbonate bags. The polycarbonate film, supplied by Nucleopore, is the stock material used for Nucleopore polycarbonate filters and is desirable because of its low trace metal content. The heat sealing is accomplished using a Teflon-coated "Quik Seal" thermal impulse sealer manufactured by National Instrument Co. The polycarbonate bags are then placed into a clean polyethylene vial (approximately 2.5 cm long $\times$ 1.0 cm ID–2/5 dram). The vial containing the coal and fly ash samples is then placed into another clean vial (5.5 cm long $\times$ 1.43 cm ID–2 dram) prior to neutron activation.

Table I.   Irradiation and Counting Procedures for Coal and Fly Ash

IRRADIATION INTERVAL/ NEUTRON FLUX*	DECAY INTERVAL	COUNT INTERVAL	INSTRUMENTATION	ELEMENTS DETERMINED
- - -	- - -	1,000-10,000 MIN	NaI (Tl) MULTI-DIMENSIONAL γ-RAY SPECTROMETER	K, Th, U
0.5-1 min/1x10^{13}	10 MIN	5 MIN	Ge (Li)	Al, Br, Cl, Cu, Mg, Mn, Na, Ti, V
6-8 hrs/5x10^{12}	4-7 DAYS	10 MIN	Ge (Li) OR ANTI-COINCIDENCE SHIELDED Ge (Li)	As, Au, Ba, Br, Ca, K, La, Lu, Na, Sm, Yb
	~30 DAYS	100-1,000 MIN		Ag, Ce, Co, Cr, Cs, Eu, Fe, Hf, Hg, Ni, Rb, Sb, Sc, Se, Sr, Ta, Tb, Th, Zn

* NEUTRONS/CM^2/SEC

Standards used for comparators are either well documented materials (U.S.G.S. BCR-1 Basalt, NBS Orchard Leaves, NBS Bovine Liver, etc.) or are prepared in the laboratory by pipetting known quantities of elements onto high purity cellulose material. Standards are weighed into polycarbonate and packaged for irradiation in the same manner as the samples.

Neutron Activation Analysis. The analysis consists of two irradiation periods and three counting intervals as shown in Table I. Gamma-ray spectrometry is accomplished by state-of-the-art Ge(Li) diodes, typically with 15–20% efficiency relative to 3 × 3″ NaI(Tl) detector and <2 KeV resolution (full width at half maximum) at the 1332 KeV ^{60}Co energy. The γ-ray spectra from samples and standards are' stored on magnetic tape and the data reduced by a PDP-15 computer. A computer program (27) was developed to identify the radionuclides present from their characteristic γ-ray energies, calculate net peak areas, and convert net peak areas to element weights by direct comparison to known standards, thus calculating the concentration of the elements in the original samples.

Table II summarizes the parameters which relate to the measurement of each element: neutron activation products, half-lives, γ-ray energies, lengths of irradiation, decay, and counting. Also listed are the possible interfering radionuclides and interfering reactions producing the same isotopes from another element which were necessarily evaluated. This table is subdivided into three sections representing the elements determined during each of the three counting intervals.

The interfering reactions were evaluated by irradiation of pure element standards with corrections applied where necessary. The only

Table II-A. Nuclear Data Relating to the Measurement of the Elements in Coal and Fly Ash

IRRADIATION PARAMETERS { IRRADIATION INTERVAL - 0.5-1.0 MIN / NEUTRON FLUX - 1×10^{13} NEUTRONS/cm^2/sec

ELEMENT	TARGET ISOTOPE	ISOTOPIC ABUNDANCE (%)	PRODUCT NUCLIDE	HALF-LIFE	THERMAL NEUTRON CROSS SECTION	BEST γ FOR MEASUREMENT (KEV)	NUMBER OF γ's PER 1000 DECAYS	ASSOCIATED γ-RAYS KEV (γ 1s/1000 DISINTEG)	POSSIBLE INTERFERING NUCLIDE IN DIODE MEASUREMENT	POSSIBLE INTERFERING NUCLEAR REACTIONS PRODUCING NUCLIDES OF INTEREST
Al	^{27}Al	100	^{28}Al	2.31m	0.24 b	1779	1000	NONE	^{151}Nd(1776)	^{28}Si(n,p) ^{28}Al / ^{31}P(n,α) ^{28}Al
Br	^{79}Br	50.5	^{80}Br	17.6m	8.5 b	617	70	666(10)	^{108}Ag(614)	^{80}Kr(n,p) ^{80}Br
Cl	^{37}Cl	24.5	^{38}Cl	37.3m	0.4 b	1643	380	2170(470)	NONE	^{38}Ar(n,p) ^{38}Cl / ^{41}K(n,α) ^{38}Cl
Cu	^{65}Cu	30.9	^{66}Cu	5.10m	2.3 b	1039	90	NONE	^{70}Ga(1040) ^{71}Zn(1040)	^{66}Zn(n,p) ^{66}Cu / ^{69}Ga(n,α) ^{66}Cu
Mg	^{26}Mg	11.2	^{27}Mg	9.45m	0.027 b	844	700	180(70)	^{101}Mo(840) ^{151}Nd(841) ^{87}Kr(846) ^{56}Mn(847)	^{27}Al(n,p) ^{27}Mg
						1014	300		^{101}Mo(1012) ^{151}Nd(1016) ^{125}Sn(1017) ^{188}Re(1018)	^{30}Si(n,α) ^{27}Mg
Mn	^{55}Mn	100	^{56}Mn	2.58h	13.3 b	847	990	2110(150)	^{27}Mg(844)	^{56}Fe(n,p) ^{56}Mn
						1811	290		NONE	^{59}Co(n,α) ^{56}Mn
Na	^{23}Na	100	^{24}Na	15.0h	0.53 b	1369	1000	2754(1000)	^{125m}Sn(1369) ^{117}Cd(1373) ^{188}Re(1368)	^{24}Mg(n,p) ^{24}Na / ^{27}Al(n,α) ^{24}Na
Ti	^{50}Ti	5.3	^{51}Ti	5.79m	0.14 b	320	950	605(15) 928(50)	^{199}Pt(317) ^{151}Nd(320) ^{105}Ru(317) ^{182m}Ta(318)	^{51}V(n,p) ^{51}Ti / ^{54}Cr(n,α) ^{51}Ti
V	^{51}V	99.8	^{52}V	3.76m	4.9 b	1434	1000	NONE	^{139}Ba(1430)	^{52}Cr(n,p) ^{52}V / ^{55}Mn(n,α) ^{52}V

Table II-B. Nuclear Data Relating to the Measurement of the Elements in Coal and Fly Ash

IRRADIATION PARAMETERS { IRRADIATION INTERVAL 6-8 HRS / NEUTRON FLUX 5×10^{12} NEUTRONS/cm^2/sec

ELEMENT	TARGET ISOTOPE	ISOTOPIC ABUNDANCE (%)	PRODUCT NUCLIDE	HALF-LIFE	THERMAL NEUTRON CROSS SECTION	BEST γ FOR MEASUREMENT (KEV)	NUMBER OF γ's PER 1000 DECAYS	ASSOCIATED γ-RAYS KEV (γ's/1000 DISINTEG)	POSSIBLE INTERFERING NUCLIDE IN DIODE MEASUREMENT	POSSIBLE INTERFERING NUCLEAR REACTIONS PRODUCING NUCLIDES OF INTEREST
As	^{75}As	100	^{76}As	26.5h	4.5 b	559	430	1220(50) 1440(7)	^{193}Os(558) ^{114m}In(558) ^{82}Br(554) ^{76}As(562) ^{122}Sb(564) ^{110m}Ag(657)	^{76}Se(n,p) ^{76}As / ^{79}Br(n,α) ^{76}As
						657	60	1789(3) 2100(3)		
Au	^{197}Au	100	^{198}Au	2.69d	98.8 b	412	980	676(10) 1088(2)	^{191}Pt(410) ^{152}Eu(411) ^{166m}Ho(411) ^{109}Cd(413) ^{177m}Lu(414) ^{109}Pd(415) ^{233}Pa(416) ^{77}Ge(416)	^{198}Hg(n,p) ^{198}Au
Ba	^{130}Ba	0.101	^{131}Ba	11.6d	8.8 b	496	480	124(280) 216(190) 373(130)	^{115m}Cd(492) ^{115}Cd(492) ^{103}Ru(492) ^{103}Pd(497)	NONE
Br	^{81}Br	49.48	^{82}Br	35.9h	3 b	777	830	554(660) 619(410) 698(270) 828(250) 1044(290) 1317(260) 1475(170)	^{187}W(772) ^{76}As(775) ^{152}Eu(779) ^{97}Mo(777) ^{131m}Te(774)	^{82}Kr(n,p) ^{82}Br / ^{85}Rb(n,α) ^{82}Br
Ca	^{46}Ca	0.0033	^{47}Ca	4.5d	0.3 b	1297	770	490(50) 815(50)	^{152}Eu(1299)	^{50}Ti(n,α) ^{47}Ca
K	^{41}K	6.77	^{42}K	12.5h	1.2 b	1524	180	310(2)	^{124}Sb(1526) ^{152}Eu(1529) ^{166m}Ho(1522)	^{42}Ca(n,p) ^{42}K / ^{45}Sc(n,α) ^{42}K
La	^{139}La	99.91	^{140}La	40.3h	8.9 b	1596	960	329(200) 487(400) 815(190) 923(100) 2530(90)	^{72}Ga(1596) ^{154}Eu(1596)	^{140}Ce(n,p) ^{140}La
Lu	^{176}Lu	2.6	^{177}Lu	6.74 d	2100 b	208	61	113(28)	^{177m}Lu(208) ^{239}Np(210) ^{121m}Te(212) ^{177m}Lu(204) ^{199}Au(208)	^{177}Hf(n,p) ^{177}Lu / ^{180}Ta(n,α) ^{177}Lu
Na	^{23}Na	100	^{24}Na	15h	0.53 b	1369	1000	2754(1000)	^{134}Cs(1365)	^{24}Mg(n,p) ^{24}Na / ^{27}Al(n,α) ^{24}Na
Sm	^{152}Sm	26.6	^{153}Sm	47.1h	210 b	103	280	70(54)	^{182}Ta(100)	^{153}Eu(n,p) ^{153}Sm / ^{156}Gd(n,α) ^{153}Sm
Yb	^{174}Yb	31.8	^{175}Yb	4.2 d	55 b	396	60	114(19) 283(37)	^{160}Tb(392) ^{67}Cu(394) ^{79}Kr(397) ^{140}La(398) ^{147}Nd(398) ^{233}Pa(398) ^{75}Se(401)	NONE

Table II-C.   Nuclear Data Relating to the Measurement of the Elements in Coal and Fly Ash

IRRADIATION PARAMETERS { IRRADIATION INTERVAL 6-8 HRS / NEUTRON FLUX 5×10^{12} NEUTRONS/cm^2/sec

ELEMENT	TARGET ISOTOPE	ISOTOPIC ABUNDANCE (%)	PRODUCT NUCLIDE	HALF-LIFE	THERMAL NEUTRON CROSS SECTION	BEST γ FOR MEASUREMENT (KEV)	NUMBER OF γ's PER 1000 DECAYS	ASSOCIATED γ-RAYS KEV (γ's/1000 DISINTEG)	POSSIBLE INTERFERING NUCLIDE IN DIODE MEASUREMENT	POSSIBLE INTERFERING NUCLEAR REACTIONS PRODUCING NUCLIDES OF INTER
Ag	^{109}Ag	48.65	^{110m}Ag	255d	3 b	658	960	680(160) 706(190) 764(230) 818(80) 885(710) 937(320) 1384(210) 1505(110)	$^{152}Eu(656)$	$^{110}Cd(n,p)^{110m}Ag$ $^{113}In(n,\alpha)^{110m}Ag$
Ce	^{140}Ce	88.48	^{141}Ce	32.5d	0.6 b	145	480	NONE	$^{233}Pa(145)$ $^{127}Xe(145)$ $^{59}Fe(142)$ $^{177m}Lu(146)$ $^{177m}Lu(147)$	$^{143}Nd(n,p)^{140}Ce$
Co	^{59}Co	100	^{60}Co	5.24y	37 b	1173 / 1332	1000 / 1000	NONE	$^{188}Re(1176)$ $^{160}Tb(1178)$ $^{79}Kr(1332)$	$^{60}Ni(n,p)^{60}Co$ $^{63}Cu(n,\alpha)^{60}Co$
Cr	^{50}Cr	4.31	^{51}Cr	27.8d	17 b	320	90	NONE	$^{192}Ir(317)$ $^{177m}Lu(319)$ $^{147}Nd(319)$ $^{193}Os(321)$ $^{177m}Lu(321)$ $^{177}Lu(321)$	$^{54}Fe(n,\alpha)^{51}Cr$
Cs	^{133}Cs	100	^{134}Cs	2.05y	30.6 b	796	990	570(230) 605(980) 1038(10) 1168(19) 1365(34)	NONE	$^{134}Ba(n,p)^{134}Cs$
Eu	^{151}Eu	47.77	^{152}Eu	12.7 y	5900 b	1408	220	122(370) 245(80) 344(270) 779(140) 965(150) 1087(120) 1113(140)	NONE	$^{152}Gd(n,p)^{152}Eu$
Fe	^{58}Fe	0.31	^{59}Fe	45 d	1.1 b	1099 / 1292	560 / 440	143(8) 192(28)	$^{160}Tb(1103)$ $^{182}Ta(1289)$ $^{115m}Cd(1290)$ $^{152}Eu(1292)$	$^{59}Co(n,p)^{59}Fe$ $^{62}Ni(n,\alpha)^{59}Fe$
Hf	^{180}Hf	35.22	^{181}Hf	43 d	10 b	482	810	133(480) 346(130)	$^{192}Ir(485)$ $^{140}La(487)$	$^{181}Ta(n,p)^{181}Hf$ $^{184}W(n,\alpha)^{181}Hf$
Hg	^{202}Hg	29.8	^{203}Hg	46.9 d	4 b	279	770	NONE	$^{197}Hg(279)$ $^{75}Se(280)$ $^{193}Os(280)$ $^{175}Yb(283)$ $^{147}Nd(275)$ $^{133}Ba(276)$	$^{203}Tl(n,p)^{203}Hg$ $^{206}Pb(n,\alpha)^{203}Hg$
Ni	^{58}Ni	67.76	^{58}Co	71 d		810	990	511(300 β$^+$) 865(14) 1670(6)	$^{125}Sb(811)$ $^{166m}Ho(811)$ $^{152}Eu(810)$ $^{154}Eu(815)$	NONE
Rb	^{85}Rb	72.15	^{86}Rb	18.6 d	1.0 b	1078	88	NONE	$^{82}Br(1081)$	$^{86}Sr(n,p)^{86}Rb$ $^{89}Y(n,\alpha)^{86}Rb$
Sb	^{123}Sb	42.75	^{124}Sb	60.2 d	3.3 b	1691	500	602(980) 723(98) 2091(69)	NONE	$^{124}Te(n,p)^{124}Sb$ $^{127}I(n,\alpha)^{124}Sb$
Sc	^{45}Sc	100	^{46}Sc	83.8d	13 b	889 / 1120	1000 / 1000	NONE	$^{110m}Ag(884)$ $^{192}Ir(884)$ $^{65}Zn(1115)$ $^{182}Ta(1121)$ $^{131m}Te(1125)$	$^{46}Ti(n,p)^{46}Sc$
Se	^{74}Se	0.87	^{75}Se	120d	30 b	265 / 280	600 / 250	97(33) 121(170) 136(570) 401(120)	$^{169}Yb(261)$ $^{115}Cd(263)$ $^{182}Ta(264)$ $^{203}Hg(279)$ $^{133}Ba(296)$ $^{192}Ir(283)$	$^{78}Kr(n,\alpha)^{75}Se$
Sr	^{84}Sr	0.566	^{85}Sr	64 d	1.4 b	514	1000	NONE	$^{85}Kr(514)$ β$^+$(511)	NONE
Ta	^{181}Ta	99.99	^{182}Ta	115d	21 b	68 / 1221	420 / 270	100(140) 152(70) 222(80) 1122(340) 1189(160) 1231(130)	$^{169}Yb(63)$ $^{121}Te(65)$ $^{75}Se(66)$ / NONE	$^{182}W(n,p)^{182}Ta$ $^{185}Re(n,\alpha)^{182}Ta$
Tb	^{159}Tb	100	^{160}Tb	72d	46 b	879	310	87(120) 299(300) 966(310) 1178(150)	$^{185}Os(879)$	$^{160}Dy(n,p)^{160}Tb$
Th	^{232}Th	100	^{233}Pa	27.0d	7.4 b	312	440	300(50)	$^{192}Ir(317)$	NONE
Zn	^{64}Zn	48.89	^{65}Zn	243d	0.46 b	1116	506	511(500 β$^+$)	$^{129m}Te(1112)$ $^{152}Eu(1112)$ $^{182}Ta(1113)$ $^{160}Tb(1115)$ $^{46}Sc(1121)$	NONE

significant interfering reactions requiring compensation proved to be $^{27}Al(n,p)^{27}Mg$ and $^{28}Si(n,p)^{28}Al$. The importance of the $^{28}Si(n,p)^{28}Al$ reaction was further minimized by use of comparator standards with a silicon content similar to that of the samples.

Irradiation I. The samples and standards (with appropriate flux monitors) are individually exposed to a neutron flux of 1×10^{13} neutrons cm^{-2} sec^{-1} for periods of ½ min for fly ash and 1 min for coal *via* a pneumatic "rabbit" system. The samples are irradiated and counted individually because of the short half-life of the (n,γ) products of some of the elements and the requirement of counting at a specific time after irradiation. A copper flux monitor is irradiated with each sample to allow adjustment for minor neutron flux variations from sample to sample. Sample sizes are approximately 10 mg for fly ash and 50–100 mg for coal.

After irradiation the primary activity is that of ^{28}Al ($t_{1/2} = 2.3$ min) which decays rapidly compared with most of the other isotopes. Figures 1 and 2 illustrate the initial activities and their decay with time. If the sample is counted immediately after irradiation, the ^{28}Al, with its higher energy photons (1779 KeV), interferes in the measurement of most other radionuclides. Conversely, if a decay interval of 20 min is allowed before counting, the ^{28}Al activity has decreased, but the disintegration rates of other radionuclides such as ^{51}Ti, ^{66}Cu, ^{52}V, and ^{27}Mg have also decayed to a point where detection is difficult. Therefore, the counting period was optimized to a 5-min counting interval beginning when the sample has been out of the reactor for 10 min. This optimization allows time

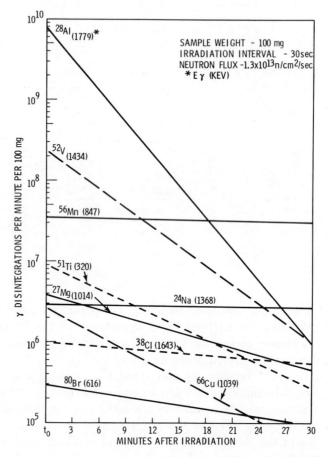

Figure 1. Concentrations of short-lived radionuclides produced by neutron activation of NBS–fly ash

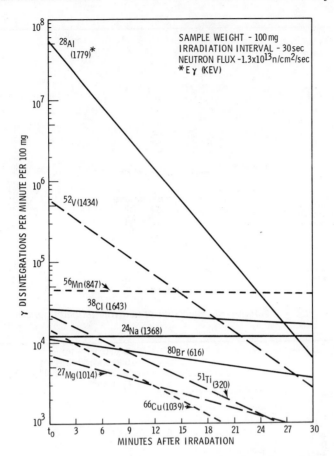

*Figure 2. Concentrations of short-lived radionuclides
produced by neutron activation of NBS–coal*

for decay of ^{28}Al activity relative to the other radionuclides but retains
sufficient activity of these other nuclides for detection by Ge(Li) spectrometry. The improvement of the activity (relative disintegration rates)
ratios of various elements to aluminum can be determined from Figure 1.
Immediately after irradiation, ^{28}Al:^{52}V is ~40:1, and after a 10-min decay
the ratio has decreased to ~10:1 and ^{28}Al:^{27}Mg has decreased from
~2000:1 to ~200:1. Figure 3 illustrates a typical γ-ray spectrum obtained after this short irradiation. In principle, multiple count intervals
after the irradiation could optimize the detection of each activation product. However, this would be prohibitively time consuming and expensive
to the research programs and does not offer a significant improvement
over the method described here.

Irradiation II.   The second irradiation involves exposure of 100-mg
samples of both coal and fly ash, along with appropriate standards, to a

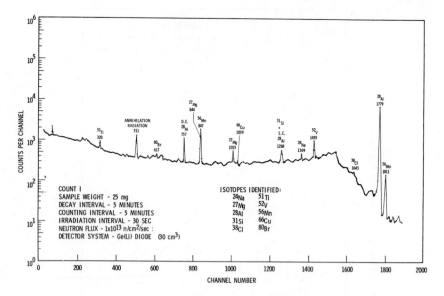

Figure 3.   Gamma-ray spectrum of neutron-activated fly ash

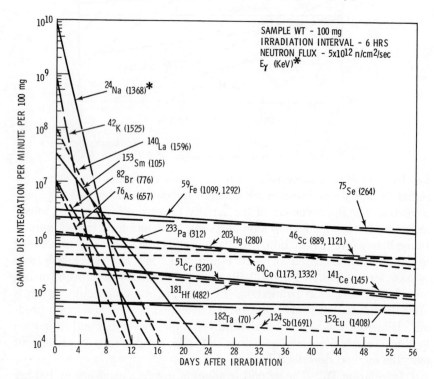

Figure 4.   Concentration of long-lived radionuclides produced by neutron activation of NBS–fly ash

neutron flux of 5×10^{12} neutron cm^{-2} sec^{-1} for 6–8 hrs followed by two counting intervals after appropriate decay periods. Figures 4 and 5 illustrate the activities of a few of the intermediate and long-lived radionuclides produced during the 6-hr irradiation and their decay with time after irradiation. These figures illustrate the presence of two groups of radionuclides with significantly different half-lives. This is the reason the samples are counted twice with each count optimized to determine one of the two groups of radionuclides. The first counting interval is for 10 min and is begun approximately 5–6 days after irradiation. At this

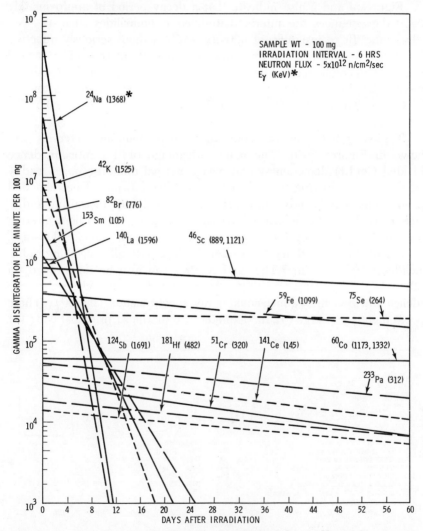

Figure 5.   Concentration of long-lived radionuclides produced by neutron activation of NBS–coal

time, as it can be seen in Figures 4 and 5, the predominant activity, ^{24}Na ($t_{1/2}$ = 15 hrs), present immediately after irradiation has decayed more than most other radionuclides of interest. The ^{24}Na, which emits a high energy γ-ray, thus no longer significantly interferes in the measurement of low energy γ-rays. Radionuclides which can be measured during the first counting period, as shown in Table I, include As, Au, Ba, Br, Ca, K, La, Lu, Na, Sm, and Yb. The determination of Na and Br from both Irradiation I and Irradiation II provides an internal check to insure consistent, accurate results.

Figures 4 and 5 also indicate that a decay period of approximately 25–30 days reduces the intermediate-lived radionuclides such as ^{24}Na, ^{140}La, and ^{82}Br to insignificant activity levels without seriously affecting the ability to measure the remaining isotopes of interest. The second count is then conducted for 100–1000 min on a Ge(Li) or anticoincidence shielded Ge(Li) spectrometer (28) and provides concentrations for Ag, Ba, Ce, Co, Cr, Cs, Eu, Fe, Hf, Hg, Ni, Rb, Sb, Sc, Se, Sr, Ta, Tb, Th, and Zn.

Typical γ-ray spectra from these two counting intervals are shown in Figures 6–10. The major advantage of the anticoincidence shielded Ge(Li) detection system over a normal diode becomes readily apparent by comparing these four figures. The anticoincidence shielded spectrometer significantly improves the sensitivity for measuring radionuclides such as ^{51}Cr or ^{65}Zn which emit a single γ-ray without altering the detection efficiency for radionuclides which emit coincident γ-rays.

Natural Radioactivity Measurement. The naturally occurring radionuclides U(^{226}Ra), Th(^{232}Th), and K(^{40}K) can also be determined on 100-g samples by direct counting of unirradiated samples using anticoincidence shielded multidimensional γ-ray spectrometers (29, 30). These spectrometers use large (12 in. diameter × 8 in. thick) principal NaI(Tl)

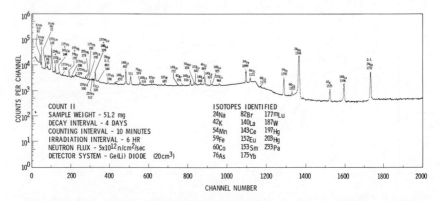

Figure 6.   Gamma-ray spectrum of neutron-activated fly ash–magnetic fraction

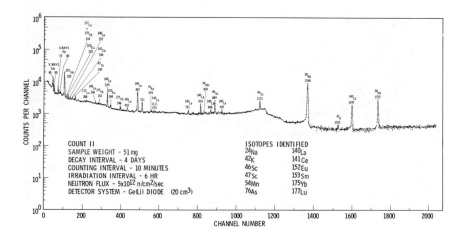

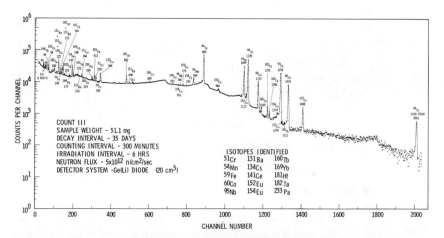

Figure 7.   (top) Gamma-ray spectrum of neutron-activated fly ash–nonmagnetic fraction. Figure 8.   (bottom) Gamma-ray spectrum of neutron-activated bottom ash–magnetic fraction.

detectors contained in an anticoincidence shield. Counting intervals of 1000–5000 min are used where very accurate measurements are desired. The analysis of multidimensional γ-ray spectra is similar to that for normal spectra (*31*), with samples being compared directly to standards prepared in a similar geometry.

This analytical method measures two elements, K and Th, which were also determined *via* NAA. However, the neutron activation procedure is limited to samples weighing a few hundred milligrams. Thus, another set of internal standards was included in the analysis to insure not only consistent and accurate results, but also homogeneous samples for these elements over a sample size range of 1000-fold.

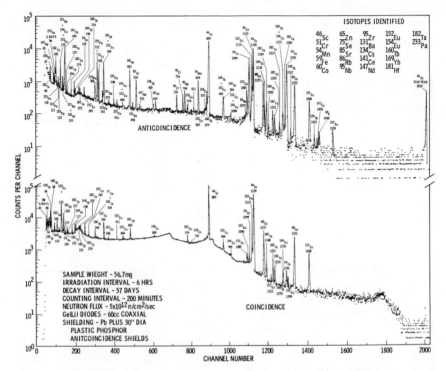

Figure 9. Anticoincidence-shielded Ge(Li) γ-ray spectra of neutron-acti-vated bottom ash

Results and Discussion

The precision for a single determination is 5–10% for most of the 40 elements in coal and fly ash. Battelle's technique of INAA has been shown to have 5–10% accuracy by the replicate analyses of well characterized standards such as U.S.G.S. Basalt Standard BCR-1 and NBS-Orchard Leaves (SRM 1571) (*32*).

An even more definitive measure of the precision and accuracy of this method is shown by the results obtained during the blind round-robin analysis of the recent NBS-EPA coal and fly ash standards. Table III compares Battelle's reported results and the NBS values for 12 elements. The overall agreement is extremely good. These samples were analyzed in Battelle's standard routine program which is currently providing thousands of high quality measurements per year for AEC programs aimed at defining the origin, transport, and ultimate deposition of atmospheric aerosols. The only unusual treatment was that six replicate samples were run. A research program using thousands of samples/yr cannot afford the luxury of many replicate samples, so the techniques must be highly

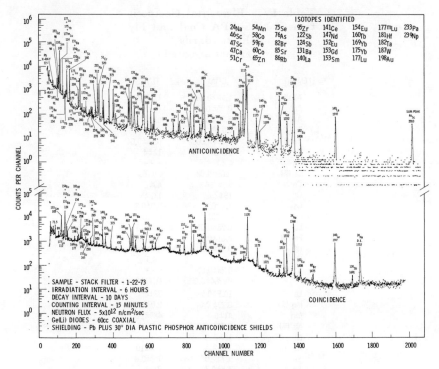

Figure 10.   Anticoincidence-shielded Ge(Li) γ-ray spectra of neutron-activated stack filter

Table III.   Comparison of NBS Certified and BNW Reported Concentrations* for NBS–EPA Fuel Standards

	FLYASH		COAL	
	NBS	BNW	NBS	BNW
As	61 ± 6	61 ± 5	5.9 ± 0.6	5.7 ± 0.5
Co	(38)	40 ± 2	---	5.2 ± 0.4
Cr	131 ± 2	131 ± 8	20.2 ± 0.5	19 ± 2
Cu	128 ± 5	< 300	18 ± 2	< 70
Fe	---	6.51%±0.31%	8700 ± 300	8100 ± 700
Hg	0.14 ± 0.01	< 0.5	0.12 ± 0.02	---
Mn	493 ± 7	489 ± 11	40 ± 3	41 ± 6
Ni	98 ± 3	---	15 ± 1	16 ± 4
Rb	(112)	124 ± 10	---	19 ± 2
Se	9.4 ± 0.5	8.8 ± 1.2	2.9 ± 0.3	3.3 ± 0.4
Sr	(1380)	1900 ± 200	---	170 ± 10
Th	(24)	28 ± 2	(3)	3.4 ± 0.6
		26.2 ± 1.3**		3.45 ± 0.10**
U	11.6 ± 0.2	---	1.4 ± 0.1	1.41 ± 0.07**
		12.0 ± 0.5**		
V	214 ± 8	220 ± 15	35 ± 3	33 ± 4

*CONCENTRATION IN PPM EXCEPT AS NOTED
( ) INDICATES INFORMATIONAL VALUE
**OBTAINED BY MULTIDIMENSIONAL GAMMA-RAY SPECTROMETRY

Table IV. Elemental Composition of NBS–EPA Coal and Fly Ash Standards
(ppm except as noted)

ELEMENT	COAL*	FLYASH*
Al (%)	1.78±0.08	12.6±0.4
Ag	0.06± 0.03	- - -
As	5.7±0.5	61±5
Au	< 0.03	- - -
Ba (%)	0.039±0.002	0.34±0.04
Br	17±2	12±4
Cd	< 2.1	- - -
Cl (%)	0.08±0.02	< 0.05
Co	5.2±0.4	40±2
Cr	19±2	131±8
Cs	1.4±0.1	9.9±0.8
Cu (%)	< 0.007	< 0.03
Eu	0.28±0.01	2.3±0.1
Fe (%)	0.81±0.07	6.51±0.31
Hf	0.97±0.10	8.2±0.8
Hg	- - -	< 0.5
K (%)	0.28±0.01	- - -
	(0.284±0.008)	(1.71±0.03)
La	10.5±0.5	82±4
Mg (%)	0.23± 0.07	2.08± 0.43
Mn	41±6	489±11
Na (%)	0.042±0.003	0.37±0.02
Ni	16±4	- - -
Rb	19± 2	124±10
Sb	3.7±2.0	7.2±0.8
Sc	3.4±0.3	27±1
Se	3.3±0.4	8.8±1.2
Sm	1.7±0.3	12.4±0.5
Sr (%)	0.017± 0.001	0.19±0.02
Ta	0.46±0.05	3.5±0.3
Tb	0.23± 0.06	2.0±0.3
Th	3.4±0.6	28±2
	(3.45±0.10)	(26.2±1.3)
Ti (%)	0.11±0.02	0.76±0.08
U	(1.41±0.07)	(12.0±0.5)
V	33±4	220±15
Yb	- - -	8.9±0.9

* AVERAGE AND STANDARD DEVIATION OF 6
DETERMINATIONS

( ) - NUMBER FROM NaI (TI) MULTIDIMENSIONAL
γ-RAY SPECTROMETRY

reliable. Table IV gives a detailed list of the other elements that have been determined in the coal and fly ash standards. These include the values from Table III as well as data on other elements for which there is no certified value from NBS. However, a comparison between this laboratory and other well qualified laboratories shows good agreement for these elements (33).

The coal and fly ash standards were certified by NBS for 250-mg sample sizes. Because of the activities induced by the neutron activation, it was necessary that the sample sizes be considerably below those recommended by NBS. Although smaller sample sizes did not affect the analytical results, an inhomogeneity of the fly ash was discovered in later measurements of magnetic and nonmagnetic fractions separated from fly ash. The difference in these two fractions becomes readily apparent in Table V. The magnetic fraction, although it constitutes only a very small percentage of the total fly ash (approximately 1%), is highly enriched in several transition metals (Fe, Mn, V, Co, Cr) and could indicate the presence of discrete metal-oxide particles. The nonmagnetic fraction is more concentrated in nonmetals such as La, Ba, Sc, Hf, Al, K, and Se, constituting the silicate fraction of the fly ash. This inhomogeneity is also evident on visual inspection of photomicrographs of the magnetic and nonmagnetic fractions. Although the presence of this inhomogeneity did not affect the analytical results for the NBS–EPA standards, it may pose a problem to other analytical techniques of high sensitivity which necessarily use smaller samples. The presence of discrete high transition metal particles may also pose an environmental hazard.

Table V. Chemical Composition *vs.* Magnetic
Characteristics of Fly Ash (ppm except as noted)

	ORIGINAL	MAGNETIC FRACTION	NON MAGNETIC FRACTION
Al (%)	10.4	9.6	17.1
Ba	500	560	750
Br	6.5	2.4	0.66
Cl (%)	<0.1	<0.1	<0.1
Co	25	50	16
Cr	116	142	116
Cs	9.5	2.9	12
Cu (%)	<0.05	<0.06	<0.05
Eu	1.3	1.1	1.4
Fe (%)	15.98	44.06	4.57
Hf	5.6	3.1	6.8
Hg	1.1	2.2	1.9
K (%)	1.93	0.82	2.44
La	47	23	56
Mn	301	1470	530
Na (%)	0.41	0.21	0.49
Rb	9	5	6
Sb	1.1	0.77	1.2
Sc	20	15	24
Se	3.2	2.1	8.8
Sm	8.6	5.8	9.5
Sr	---	---	560
Ta	2.0	1.3	2.7
Tb	1.3	0.84	1.5
Th	15	13	19
Ti (%)	0.57	1.72	2.42
V	157	721	514

Although the NBS standards were collected from several sources, some tentative conclusions can be reached concerning the combustion process. Using selected elements for comparison which are of a refractory nature and would be expected to be conserved in the ash following the combustion process, it becomes apparent that other elements such as selenium and antimony are not being conserved, but rather are lost from the ash. Refractory elements selected for illustration are Eu, Hf, Sc, Ta, Tb, and La, although most other elements display the same conservative behavior. Table VI shows the concentrations of these elements in both

Table VI. Coal-to-Fly Ash Ratios for Selected Elements in NBS–EPA Standards

	COAL	FLYASH	COAL/FLYASH
Eu	0.28	2.3	0.12
Hf	0.97	8.2	0.12
Sc	3.4	27	0.13
Ta	0.46	3.5	0.13
Tb	0.23	2.0	0.12
La	10.5	82	0.13

CONCENTRATIONS IN PARTS PER MILLION

the coal and fly ash and the coal-to-fly ash ratios. If these elements are totally conserved in the ash, this ratio is synonomous with the ash content of the original coal. The coal-to-fly ash ratio for all of these elements is nearly identical and averages 1:7 which indicates a reasonable ash content of 12.5% for the coal.

Selenium and antimony, by contrast, behave quite differently. Using the 12.5% ash content from above, one would expect that the elements would be concentrated in the fly ash by a factor of 8.3. However, these elements are not, and only approximately 30% and 20% of the selenium and antimony, respectively, appear to be present in the fly ash.

Secondly, it is possible to draw conclusions concerning the source of certain elements found in the coal. By normalizing Eu, Hf, Sc, Ta, and Tb to La in coal and comparing these with similar normalizations for various crustal components (34) as shown in Table VII, it appears that, for these elements, the crustal average and coal composition are quite similar. This is expected because these elements are generally not concentrated in organic matter, but rather should be associated with geological materials intermixed with the coal. These geological materials should approximate average crustal composition because they are various weathered materials, sands, silts, etc., which have been sedimented with the coal body.

Table VII   Ratios to Lanthanum for Selected Elements in
NBS–EPA Coal Standard and Other Crustal Materials

	COAL	GRANITE*	CRUSTAL AVERAGE*	DIABASE*
Eu	2.7×10^{-2}	8.3×10^{-3}	4×10^{-2}	3.7×10^{-2}
Hf	9.2×10^{-2}	4.3×10^{-2}	1×10^{-1}	$5. \times 10^{-2}$
Sc	3.2×10^{-1}	2.5×10^{-2}	7.3×10^{-1}	1.1
Ta	4.4×10^{-2}	1.3×10^{-2}	6.7×10^{-2}	2.3×10^{-2}
Tb	2.2×10^{-2}	9.2×10^{-3}	3×10^{-2}	2×10^{-2}

* MASON (1966)

A similar normalization for the coal and various crustal components can be performed using the transition elements Co, Cr, Fe, Mn, Ni, and V. Table VIII shows this normalization to iron for coal, granite, diabase, and crustal average materials. As with the earlier elements, the ratio of transition metals appears to approximate the crustal average composition. Thus, it appears that the concentrations of the transition elements may also be explained by the incorporation of geological materials in the coal.

Many of the elemental concentrations in the coal may be thus attributed to the presence of intermixed geological materials. This agrees with other findings recently reported by Ruch *et al.* (35). Therefore, emissions from coal combustion could be significantly reduced by cleansing the coal of these intermixed geological materials prior to combustion.

Summary

A sensitive and accurate INAA method for coal and fly ash has been developed which can simultaneously determine approximately 40 elements consisting of major, minor, and trace constituents. This multi-

Table VIII.   Ratios to Iron for Selected Transition Metals
in NBS–EPA Coal Standard and Other Crustal Materials

	COAL	GRANITE*	CRUSTAL AVERAGE*	DIABASE*
Co	6.4×10^{-4}	1.8×10^{-4}	5×10^{-4}	6.4×10^{-4}
Cr	2.3×10^{-3}	1.6×10^{-3}	2×10^{-3}	1.5×10^{-3}
Mn	5.1×10^{-3}	1.7×10^{-2}	1.9×10^{-2}	1.7×10^{-2}
Ni	2.0×10^{-3}	1.5×10^{-4}	1.5×10^{-3}	1.0×10^{-3}
V	4.1×10^{-3}	1.2×10^{-3}	2.7×10^{-3}	3.1×10^{-3}

* MASON (1966)

element technique facilitates the use of elemental ratios and coal-to-fly ash ratios. The use of these ratios allow conclusions to be drawn concerning the origin of many trace elements in coal and their fate during combustion. This technique is currently being used with excellent results in studies of the effluent and ash residues from coal-fired electric generating plants. The goals of this ongoing work include characterizing the emissions from fossil fuel facilities and defining subsequent atmospheric transport and environmental cycling processes. Battelle thus hopes to evaluate the potential environmental impact of the trace elements from combustion of fossil fuels, a facet of fossil fuel energy generation that has been largely neglected during previous studies of environmental impact.

Acknowledgment

The authors wish to acknowledge and thank W. L. Butcher, J. H. Reeves, and C. L. Wilkerson of this laboratory for providing invaluable analysis of samples for this study. W. C. Weimer is thanked for his editorial comments.

Literature Cited

1. North Central Power Study (NCPP), Report of Plan 1, I and II, October, 1971.
2. Dept. of the Interior, U.S. Energy, A Summary Review (1972).
3. Johnson, N. M., Reynolds, R. C., Likens, G. E., "Atmospheric Sulfur: Its Effect on the Chemical Weathering of New England," Science (1972) 177, 514–516.
4. Likens, G. E., Bormann, F. H., "Acid Rain: A Serious Regional Environmental Problem," Science (1974) 184, 1176–1179.
5. Lundholm, B., in "Global Effects of Environmental Pollution," S. F. Singer (Ed.), pp. 195–201, Springer–Verlag, New York, 1970.
6. Rancitelli, L. A., unpublished data.
7. Eisenbud, M., Petrow, H. G., "Radioactivity in the Atmospheric Effluents of Power Plants that Use Fossil Fuels," Science (1964) 144, 288–289.
8. Bedrosian, P. H., unpublished data (1969).
9. Martin, J. E., Harward, E. D., Oakley, D. T., "Comparison of Radioactivity from Fossil Fuel and Nuclear Power Plants, Environmental Effects of Producing Power," Part I, Appendix 14, Committee Print, Joint Committee on Atomic Energy, U.S. Congress (91st), (1969).
10. Billings, C. E., Sacco, A. M., Matson, W. R., Griffin, R. M., Coniglio, W. R., Harley, R. A., "Mercury Balance on a Large Pulverized Coal-Fired Furnace," J. Air Pollut. Contr. Ass. (September 1973) 23, (9), 773–777.
11. Diehl, R. C., Hattman, E. A., Schultz, H., Haren, R. J., "Fate of Trace Mercury in the Combustion of Coal," U.S. Dept. of the Interior, Bureau of Mines Managing Coal Wastes and Pollution Program, Technical Progress Report, 54, May 1972.
12. Fancher, J. R., "Trace Elements Emissions from the Combustion of Fossil Fuels," Environmental Resources Conference on Cycling and Control of Metals, Battelle—Columbus, Ohio, November 1, 1972.

13. Natusch, D. F. S., Wallace, J. R., Evans, Jr., C. A., "Toxic Trace Elements: Preferential Concentration in Respirable Particles," *Science* (January 1974) **183**.
14. Rancitelli, L. A., unpublished data.
15. Rancitelli, L. A., Cooper, J. A., Perkins, R. W., "Multi-element Characterization of Atmospheric Aerosols by Neutron Activation and Direct Gamma Ray Analysis and X-ray Fluorescence Analysis," IAEA Symposium on Nuclear Techniques in Comparative Studies of Food and Environmental Contamination, Otaniemi, Finland, August 1973, **BNWL-SA-4671**, Battelle-Northwest, Richland, Wash., 1973.
16. Tanner, T. M., Young, J. A., Cooper, J. A., unpublished data (1974).
17. Tanner, T. M., Rancitelli, L. A., unpublished data.
18. Peirson, D. H., Cowse, P. A., Salmon, L., Cambray, R. S., "Trace Elements in the Atmospheric Environment," *Nature* (January 1973) **241**, 252–256.
19. John, W., Kaifer, R., Rahn, K., Wesolowski, J. J., "Trace Element Concentrations in Aerosols from the San Francisco Bay Area," *Atmos. Environ.* (1973) **7**, 107–118.
20. Bogen, J., "Trace Elements in Atmospheric Aerosol in the Heidelberg Area, Measured by Instrumental Neutron Activation Analysis," *Atmos. Environ.* (1973) **7**, 1117–1125.
21. Gordon, G. E., Zoller, W. H., Gladney, E. S., "Abnormally Enriched Trace Elements in the Atmosphere," Annual Conference on Trace Substances in Environmental Health, **7**, Univ. of Missouri, Columbia, Mo., June, 1973.
22. Gladney, E. S., Zoller, W. H., Jones, A. G., Gordon, G. E., "Composition and Size Distributions of Atmospheric Particulate Matter in Boston Area," *Environ. Sci. Technol.* (June 1974) **8**, 551–557.
23. Harrison, P. R., Rahn, K. A., Dams, R., Robbins, J. A., Winchester, J. W., "Areawide Trace Metal Concentrations Measured by Multielement Neutron Activation Analysis: A One Day Study in Northwest Indiana," *APCA J.* (September 1971) **21**, (9), 563–570.
24. Zoller, W. H., Gladney, E. S., Duce, R. A., "Atmospheric Concentrations and Sources of Trace Metals at the South Pole," *Science* (January 1974) **183**, 198–200.
25. Rahn, K. A., "Sources of Trace Elements in Aerosols—An Approach to Clean Air," Technical Report, Univ. of Michigan, College of Engineering, Dept. of Meteorology and Oceanography, May 1971.
26. Rancitelli, L. A., Perkins, R. W., "Trace Element Concentration in the Troposphere and Lower Stratosphere," *J. Geophys. Res.* (1970) **75**, 3055–3064.
27. Kosorok, J. R., *IEEE Trans. Nucl. Sci.* (1972) **NS-19** (1), 640.
28. Cooper, J. A., Perkins, R. W., "A Versatile Ge(Li)–NaI(Tl) Coincidence–Anticoincidence Gamma-Ray Spectrometer for Environmental and Biological Problems," *Nucl. Instrum. Methods* (1969) **74**, 197–212.
29. Wogman, N. A., Perkins, R. W., Kaye, J. H., "An All Sodium Iodide Anticoincidence Shielded Multidimensional Gamma-Ray Spectrometer for Low-Activity Samples," *Nucl. Instrum. Methods* (1969) **74**, 197–212.
30. Perkins, R. W., "An Anticoincidence Shielded-Multidimensional Gamma-Ray Spectrometer," *Nucl. Instrum. Methods* (1965) **33**, 71–76.
31. Perkins, R. W., Rancitelli, L. A., Cooper, J. A., Kaye, J. H., Wogman, N. A., "Cosmogenic and Primordial Radionuclide Measurements in Apollo 11 Lunar Samples by Nondestructive Techniques," *Proc. Apollo 11 Lunar Sci. Conf.* (1970) **2**, 1455–1469.
32. Robertson, D. E., Rancitelli, L. A., Langford, J. C., Perkins, R. W., IDOE Baseline Studies of Pollutants in the Marine Environment and Research Recommendations, 1972.

33. Ondov, J. M., Zoller, W. H., Olmez, I., Aras, N. K., Gordon, G. E., Ranci-
    telli, L. A., Abel, K. H., Filby, R. H., Shah, K. R., Ragaini, R. C., un-
    published data (1974).
34. Mason, B., "Principles of Geochemistry," 3rd ed., John Wiley and Sons,
    New York, 1966.
35. Ruch, R. R., Gluskoter, H. J., Shimp, N. F., Ill. State Geological Survey
    Report #61, April, 1973.

RECEIVED July 12, 1974. This paper is based on work performed for the United
States Atomic Energy Commission Contract AT(45-1)-1830.

The Fate of Some Trace Elements During Coal Pretreatment and Combustion

HYMAN SCHULTZ, E. A. HATTMAN, and W. B. BOOHER

U.S. Department of the Interior, Bureau of Mines, 4800 Forbes Ave., Pittsburgh, Pa. 15213

Preliminary studies have shown that it is possible to remove over half of the potentially toxic trace elements present in coal when the mineral matter is reduced by coal washing. When coal is burned in a power plant, about 13% of the mercury and about 50% of the lead and cadmium may remain with the fly ash. Analytical chemical techniques have been developed to determine Hg, Cu, Cr, Mn, Ni, Cd, Pb, and F in coal and fly ash. These techniques produce accurate and precise results despite the fact that there are no coals with established trace element content, except for mercury.

The fact that coal contains many potentially toxic trace elements, together with the fact that over 300 million tons of coal were burned in the United States in 1972 to generate electric power, has led to a great deal of interest in the trace element content of coal. Questions still to be answered include the following: Exactly what quantities of potentially toxic trace elements exist in the coal we mine? About 40% of the coal burned to generate electricity is washed (1). What is the effect of washing on the trace element content of the coal? Exactly how much of these elements enters the environment *via* the stacks of coal-burning power plants?

Two trace element studies are being done to answer some of these questions. The first study, funded by the Environmental Protection Agency, concerns coal washing and is designed to determine the distribution of certain trace elements in coal. The various specific gravity fractions of coal are separated by a sink–float procedure. Commercially available organic fluids are used as the separation media.

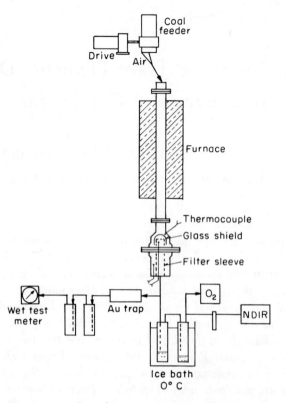

Figure 1.   Flow diagram of a trace element unit

The procedure used insures that the coal samples analyzed are representative of the seam being sampled. A 600-lb channel sample is obtained from as near the working face of the mine as possible. The coal sample is coned, long piled, and shoveled into four pans using ASTM procedures. The original sample is divided in half by combining opposite pans. One of the halves of the original sample is crushed and riffled in six separate stages until a 3-lb sample is obtained. The 3-lb sample, which is used for the specific gravity separation, will pass through a 14 mesh screen (0.046″).

The second program, funded by the U.S. Bureau of Mines, is concerned with the fate of the various toxic trace elements present when coal is burned in power plants. Coals and ashes from experimental combustors and power plants are collected and chemically analyzed. Comparing the amount of a trace element in a coal with the amount found in the ash resulting from the combustion of that coal allows us to determine the maximum amount of that element that could be emitted into the environment *via* the power plant stacks.

The initial experiments for each element of interest are usually conducted on the 100-gram-per-hour laboratory combustor shown in Figure 1. The boxes labeled O_2 and NDIR in Figure 1 stand for "oxygen analyzer" and "nondispersive infrared analyzer," respectively, and were originally intended to provide flue gases analyses during each experiment. Difficulties with the equipment operation prevented the use of either the oxygen analyzer or the NDIR during the experiments described in this report. The gases are analyzed by withdrawing samples and using gas chromatography. The O_2 and NDIR are included in the diagram for completeness. This combustor uses crushed coal (100–200 mesh) and features free fall combustion and Nomex bag ash collection. Gas flow rates of 10 standard cubic ft/hr (SCFH) for primary air and 30 SCFH for secondary air are used. A small combustor allows greater control and sample definition and helps to bring out problems while the work is still on a small scale.

The next experiments are conducted on a 500-pound-per-hour combustor. This combustor, which simulates commercial practice, is shown in Figure 2. It is a wall-fired, dry-bottom unit which uses cyclone ash collection. Where possible both coal and ash samples are obtained from commercial power plants to verify the results from the experimental combustors. The combination of the two programs will hopefully give

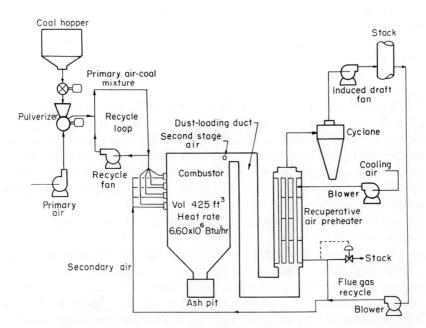

Figure 2. Simplified flowsheet of 500-lb/hr pulverized-fuel-fired furnace

us a picture of the fate of some trace elements from the coal mine to the power plant stack.

This paper covers the problems encountered and the results obtained to date in both programs. The results that are presented are of a preliminary nature because both programs are still active. The elements that are currently being studied in the EPA program are mercury, copper, chromium, manganese, nickel, cadmium, lead, and fluorine. In the Bureau program, the studies have been limited to mercury, cadmium, and lead.

Contamination is a problem one always faces in trace analysis. Mercury is ubiquitous in many laboratories, as is fluoride. Lead is present in dust, particularly in laboratories located close to heavy automobile traffic. Extreme caution must always be exercised as contamination on the trace level may come from unexpected sources. Table I shows that using manganese steel in the jaws used to crush the coal increased the manganese content of the crushed coal more than twofold.

Table I. Analysis of Coal Before and After Crushing

Element	Concentration Before Crushing (ppm)	Concentration After Crushing (ppm)
Cr	3.4	4.3
Cu	3.2	5.4
Mn	4.1	10.3

Another problem encountered when analyzing coal for trace elements is the lack of standards. Except for mercury, there is no certified standard coal presently available for use in trace element analysis. Because the precision and accuracy of the analytical procedures used are in many cases affected by the matrix one is dealing with, the lack of a standard coal is a serious problem.

To deal with the matrix problem, the method of standard additions is used, in which the sample itself serves as the matrix for preparing "knowns." Known amounts of the analyte are added to aliquots of the sample, and these are used to construct the working curve. When a linear relationship exists between the analytical effect being measured (absorbance, for instance) and the analyte concentration, this procedure produces a working curve with a negative intercept. The magnitude of the negative intercept along the concentration axis will be equal to the concentration of the analyte in the sample. To check the precision analyses are always replicated and the data accumulated for statistical purposes.

Material balances are done where possible to check overall precision and accuracy. In the coal-washing study, material balances are always possible. In cases where enough data for statistical calculations has not been accumulated, an arbitrary limit of ±15% for acceptable material balances has been set.

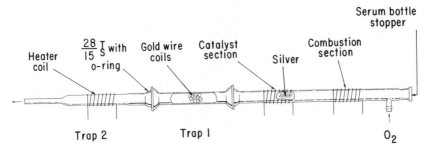

Figure 3. Double-gold amalgamation train

Analytical Methods

Since the details of the analytical methods are to be published in the near future, they are only outlined here.

Mercury. A double-gold amalgamation–flameless atomic absorption procedure was selected for mercury determination after a study of the available analytical techniques (2). The double-gold amalgamation train is shown in Figure 3. Because mercury is volatile and can be quantitatively separated from the coal matrix, calibration can be accomplished with mercury-saturated air. Figure 4 shows mercury-saturated air being injected into the double-gold amalgamation train. Figure 5 is a diagram of the mercury-saturated air container. The response obtained for various

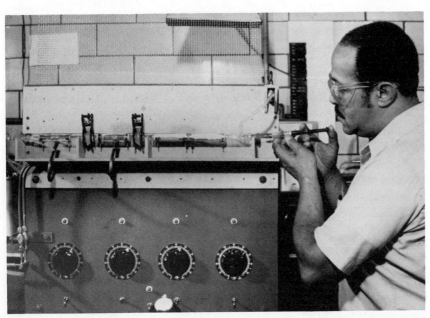

Figure 4. Injection of mercury-saturated air

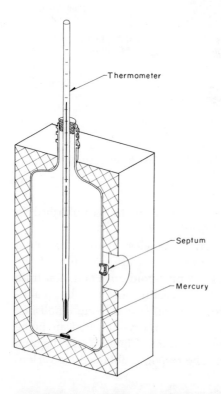

Figure 5. Calibration container for mercury-saturated air

amounts of mercury-saturated air is shown in Figure 6. This trace was obtained with minimum instrumental amplification. Mercury-saturated air is normally used for calibration purposes on a day-to-day basis and NBS standard 1630s periodically analyzed to check our procedure. Table II presents the results obtained with the NBS standard coal through March 2, 1973.

Fluorine. The fluorine determination was adapted from an older method (*3*). Fluorine in coal is determined by combusting the coal in the presence of calcium oxide, fusing the residue with sodium carbonate, leaching the melt with phosphoric acid, distilling the solution with sulfuric acid, concentrating the fluoride by passing the solution through an anion exchange resin (Amberlite IRA 410), and determining the fluoride content of the solution with a fluoride specific ion electrode using the standard additions method. An NBS opal glass standard is frequently carried through the entire procedure to check fluorine recovery.

Certain precautions are observed in the fluoride determination to deal with contamination problems. The reagents and deionized water used are kept separate and used only for the fluoride analysis. Each new batch of water or reagent is analyzed for fluoride. Finally, the fluoride analysis is carried out in a laboratory where hydrofluoric acid has never been used.

Copper, Chromium, Manganese, and Nickel. The analytical method for determining copper, chromium, manganese, and nickel involves digesting the coal with nitric and perchloric acids, fusing the residue with lithium metaborate, and determining the combined digestion and leach solutions by atomic absorption spectrophotometry. Since there is no standard material to analyze for the construction of calibration curves, the standard additions method is used for the assay. While this method increases the time required for analysis, it helps to eliminate the effect of the matrix.

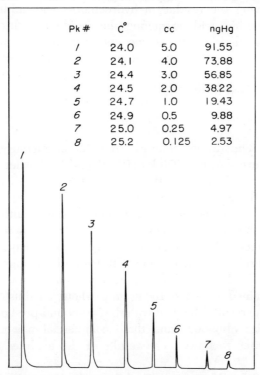

Pk #	c°	cc	ngHg
1	24.0	5.0	91.55
2	24.1	4.0	73.88
3	24.4	3.0	56.85
4	24.5	2.0	38.22
5	24.7	1.0	19.43
6	24.9	0.5	9.88
7	25.0	0.25	4.97
8	25.2	0.125	2.53

Figure 6.   Response for mercury-saturated air injections

Table II.   Analysis of NBS SRM No. 1630[a]

Dates	Number of Replicates	Bureau Value
2/24/72	5	0.12 ±0.02
4/19/72	5	0.12 ±0.01
7/5/72	6	0.14 ±0.02
9/24–12/13/72	39	0.13 ±0.03
2/1–3/2/73	37	0.13 ±0.02

[a] NBS certified mercury value = 0.126 ± 0.006 ppm Hg.

One of the contamination problems that was encountered illustrates the types of problems that occur in this work. A consistently high material balance was being obtained for copper. One of the platinum crucibles had been used before in the analysis of copper-containing materials. Vigorous cleaning procedures had not removed all of the copper. Now, separate virgin platinum crucibles are used for the trace work and contamination from this source has been eliminated.

Cadmium and Lead. Cadmium and lead are determined in coal by ashing the coal at 500°C, treating with hydrofluoric and hydrochloric acids, fusing the residue with potassium carbonate, evaporating to near dryness, dissolving the residue in hydrochloric acid, adding potassium iodide and ascorbic acid, extracting the iodide complex of lead and cadmium into methyl isobutyl ketone, and aspirating the methyl isobutyl ketone layer into the flame of an atomic absorption spectrophotometer. Blanks are carried through the entire procedure, and the standard additions method is used for calibration.

Results

Mercury. The results obtained to date for mercury in the coal-washing study are shown in Tables III, IV, and V. Earlier results of this study have been published by the U.S. Bureau of Mines (4). Removal of that part of the coal having a specific gravity greater than 1.60 would reduce both the sulfur and the mercury contents significantly in each case.

Table VI shows the analysis of the coals used in the combustion studies for mercury. P-3 is a Tebo and Weir seam mixture from Henry County, Missouri while DRB-E and MR are both Pittsburgh seam coals originating in Washington County, Pennsylvania. Table VII presents the results obtained with the 100-gram-per-hour combustor, and Table VIII presents the results obtained with the 500-pound-per-hour combustor.

Ash samples obtained from three commercial power plants were analyzed for trace mercury. The results of the analyses are shown in

Table III. Mercury in Coal from the Upper Freeport Coalbed, Garrett County, Md.

Specific Gravity Fraction	% Coal in Fraction	Mercury Concentration (µg/g)
Head coal	100	0.28 ± 0.02
FL. 1.30	37.6	0.08 ± 0.02
FL. 1.30–1.40	36.7	0.16 ± 0.03
FL. 1.40–1.60	10.3	0.56 ± 0.06
Sink 1.60	15.4	1.13 ± 0.03

Reduction in concentration on removal of sink 1.60 fraction: mercury 47%; sulfur 42%.

Table IV.   Mercury in Strip Coal from the Indiana V Coalbed, Pike County, Ind.

Specific Gravity Fraction	% Coal in Fraction	Mercury Concentration ($\mu g/g$)
Head coal	100	0.13
FL. 1.30	42.6	0.09 ± 0.03
FL. 1.30–1.40	36.3	0.08 ± 0.03
FL. 1.40–1.60	13.7	0.15 ± 0.03
Sink 1.60	7.4	0.59 ± 0.05

Reduction in concentration on removal of sink 1.60 fraction: mercury 28%; sulfur 42%.

Table V.   Mercury in Coal from the Lower Kittanning Coalbed, Centre County, Md.

Specific Gravity Fraction	% Coal in Fraction	Mercury Concentration ($\mu g/g$)
Head coal	100	0.26
FL. 1.30	20	0.16
FL. 1.30–1.40	28.1	0.23
FL. 1.40–1.60	24.8	0.19
Sink 1.60	27.1	0.43

Reduction in concentration on removal of sink 1.60 fraction: mercury 24%; sulfur 23%.

Table VI.   Analysis of Combusted Coals[a]

	DRB-E	P-3	MR
Proximate analysis			
Volatile matter (%)	35.8	37.8	37.7
Fixed carbon (%)	57.3	40.6	52.2
Ash (%)	6.9	21.6	10.1
Ultimate analysis			
Hydrogen (%)	5.1	4.4	5.0
Carbon (%)	78.1	61.9	74.2
Nitrogen (%)	1.6	1.0	1.5
Oxygen (%)	7.1	5.9	7.1
Sulfur (%)	1.2	5.2	2.1
Ash (%)	6.9	21.6	10.1
Calorific value (Btu)	13,970	11,190	13,310
Free swelling index	8	2.5	1.5

[a] On moisture free basis.

Table VII. Summary of Results from

Coal-Run Number	Coal Feed Rate (g/hr)	Combustion Efficiency (%)	Fly Ash Production Rate (g/hr)
DRB-E–1	98.1	97.6	9.1
DRB-E–2	105.1	97.0	10.4
DRB-E–3	108.1	96.4	11.3
P-3–1	98.9	97.5	23.8
P-3–2	135.7	96.3	34.3
P-3–3	117.2	98.3	27.1

[a] Average value of 12 replicates for DRB-E.
[b] Average value of 21 replicates for P-3.
[c] No flue gas sampling on this run.

Table VIII.   Summary of Results from a
500-pound-per-hour Combustor

Mercury in Coal (MR) (μg Hg/g coal)	Number of Replicates	Mercury in Fly Ash (μg Hg/g ash)	Number of Replicates	% of Total Mercury Found in Fly Ash
0.18 ± 0.04	23	0.22 ± 0.04	17	12 ± 3

Table IX.   Summary

	Steam Conditions		
Type of Firing	Rate (10^3 lb/hr)	Pressure (psig)	Temperature (°F)
Slag top, cyclone	702	1268	899
Slag top, pulverized coal	723	1293	799
Dry bottom, tangential	461	1450	928

Table X.   Fluorine in Coal from the Lower Kittanning
Coalbed, Centre County, Penn.

Specific Gravity Fraction	% Coal in Fraction	Fluorine Concentration (μg/g)
Head coal	100	137
FL. 1.30	20.1	30
FL. 1.30–1.40	30.3	56
FL. 1.40–1.60	24.0	123
Sink 1.60	25.6	270

Reduction in concentration on removal of sink 1.60 fraction: fluorine 42%;
sulfur 23%.

a 100-gram-per-hour Combustor

Coal ($\mu g/g$)	Mercury Content ($\pm$ one standard deviation)		% of Total Mercury in Fly Ash	% of Total Mercury Accounted for
	Fly Ash ($\mu g/g$)	Flue Gas ($\mu g/m^3$)		
	0.97 ± 0.05	2.2	60	77
0.15 ± 0.02[a]	0.83 ± 0.13	6.5	55	101
	0.95 ± 0.09	1.7	66	78
	0.31 ± 0.04	7.4	31	62
0.24 ± 0.05[b]	0.35 ± 0.06	—[c]	37	—[c]
	0.37 ± 0.04	14.4	36	94

Table IX. Using the published values (5) for the mercury content of Illinois No. 6 HVCB coal, the percentage of the mercury in the burned coal that was retained by the ash is not significantly different from the values found with the 500-pound-per-hour combustor (*i.e.*, 13% *vs.* 12% retained). This supports the contention that the 500-pound-per-hour combustor simulates commercial practice. On the basis of these findings, it appears that the maximum percentage of mercury that could be emitted

of Power Plant Data

Coal Fired	Fly Ash Sample Collection Conditions	Mercury Content of Fly Ash ($\mu g\ Hg/g$)
Illinois #6 Hvcb	mechanical collector hopper	0.10 ± 0.02
Illinois #6 Hvcb	electrostatic precipitator hopper	0.26 ± 0.04
Kentucky #6 Hvbb	mechanical dust collector hopper	0.22 ± 0.02

Table XI. Fluorine in Coal from an Uncorrelated Coalbed, Mahoska County, Iowa

Specific Gravity Fraction	% Coal in Fraction	Fluorine Concentration ($\mu g/g$)
Head coal	100.0	100
FL. 1.30	20.5	65
FL. 1.30–1.40	30.3	85
FL. 1.40–1.60	22.0	114
Sink 1.60	27.2	110

Reduction in concentration on removal of sink 1.60 fraction: fluorine 6%; sulfur 50%.

from the stacks of a coal-burning power plant is roughly 85–90%. This result has been verified by other studies (6, 7, 8).

Fluorine. Tables X, XI, and XII show the results obtained for fluorine in the coal-washing study with coals from Pennsylvania, Iowa, and West Virginia. In all three coals, the fluorine appears to be associated with the mineral matter in the coal. Therefore, removal of the higher specific gravity fractions would lower the fluoride content of the remaining coal.

Chromium, Nickel, Copper, and Manganese. Tables XIII and XIV show the results obtained in the coal-washing study for chromium, nickel, copper, and manganese with a Maryland coal and a Kentucky coal. For the reasons stated earlier, the manganese results should be viewed with caution. Tables XIII and XIV also show the reductions in concentration of these trace elements and of sulfur that could be achieved by removing the highest specific gravity fraction of the coal. These elements are more concentrated in the mineral matter than in the organic fraction of the coal.

Lead and Cadmium. Table XV shows the results obtained in the coal-washing study for cadmium and lead in a Hazard No. 4 coal from

Table XII. Fluorine in Coal from the Stockton-Lewiston Coalbed, Amhersdale, W. Va.

Specific Gravity Fraction	% Coal in Fraction	Fluorine Concentration ($\mu g/g$)
Head coal	100	71
FL. 1.30	30.6	8
FL. 1.30–1.40	34.0	44
FL. 1.40–1.60	18.9	123
Sink 1.60	16.5	155

Reduction in concentration on removal of sink 1.60 fraction: fluorine 31%; sulfur rises 7%.

Table XIII. Cr, Cu, Ni, and Mn in Coal from the Upper Freeport Coalbed, Garrett County, Md.

Specific Gravity Fraction	% Coal in Fraction	Concentration ($\mu g/g$)			
		Cr	Cu	Ni	Mn
Head coal	100	27	16	16	13
FL. 1.30	37.6	13	7.0	8.1	2.5
FL. 1.30–1.40	36.7	23	8.8	9.2	6.5
FL. 1.40–1.60	10.3	34	24	26	23
Sink 1.60	15.4	73	58	38	51

Reduction in concentration on removal of sink 1.60 fraction: Cr 29%; Cu 43%; Ni 28%; Mn 50%; S 42%.

Table XIV. Cr, Cu, Ni, and Mn in Coal from the Hazard No. 4
Coalbed, Bell County, Ky.

Specific Gravity Fraction	% Coal in Fraction	Concentration ($\mu g/g$)			
		Cr	Cu	Ni	Mn
Head coal	100	26	28	18	263
FL. 1.30	51.1	6	13	10	30
FL. 1.30–1.40	16.9	11	26	15	89
FL. 1.40–1.60	9.2	33	55	28	240
Sink 1.60	22.9	73	66	38	967

Reduction in concentration on removal of sink 1.60 fraction: Cr 56%; Cu 29%; Ni 30%; Mn 75%; S 18%.

Table XV. Lead and Cadmium in Coal from the Hazard No. 4
Coalbed, Bell County, Ky.

Specific Gravity Fraction	% Coal in Fraction	Concentration ($\mu g/g$)	
		Cadmium	Lead
Head coal	100	0.12	14
FL. 1.30	51.0	0.08	4
FL. 1.30–1.40	16.9	0.20	10
FL. 1.40–1.60	9.2	0.24	25
Sink 1.60	22.9	0.10	40

Reduction in concentration on removal of sink 1.60 fraction: cadmium 0%; lead 56%; sulfur 18%.

Table XVI. Lead and Cadmium in a Pittsburgh Seam Coal (±1 S.E.)[a]

Lead Content (ppm)	Number of Replicates	Cadmium Content	Number of Replicates
7.7 ± 0.5	9	0.14 ± 0.05	9

[a] Deep mined in Washington County, Penn.

Kentucky. The cadmium appears to be so distributed in the coal that removal of the high specific gravity fraction of the coal does not affect the cadmium concentration in the remaining coal. This result is different from that encountered with the other trace elements. Other coals will be studied to see if they exhibit the same cadmium distribution as the Hazard No. 4 coal.

Table XVI shows the lead and cadmium content of one of the Pittsburgh seam coals used in the coal combustion study, and Table XVII shows the results obtained when the coal was combusted in the two experimental furnaces. Compared with mercury, a greater amount of both the cadmium and the lead were retained by the ash in both combustors.

It appears that the amount of a trace element retained in the fly ash depends upon the vapor pressure of that element at combustion temperature. There are undoubtedly other factors involved, and some of them will be elucidated in future studies.

Table XVII.   Lead and Cadmium in Fly Ash

	Cadmium		Lead	
Combustor	in Ash (ppm)	Accounted for (%)	in Ash (ppm)	Accounted for (%)
100 g/hr	1.0	68	—	—
100 g/hr	0.74	54	71	92
100 g/hr	0.99	76	68	93
500 lb/hr	1.22	101	49	72
500 lb/hr	0.78	65	44	64
500 lb/hr	0.36	37	25	46

Summary and Conclusions

Although these findings are part of a continuing program and are somewhat preliminary in nature, certain conclusions can be drawn from them. First, one must conclude that for those trace elements studied, except possibly cadmium, elimination of the higher specific gravity fractions will reduce the trace element content of coal. Since coal washing is used to lower the ash and sulfur content of coals, the accompanying reduction of the trace element content is an added benefit. The effects of coal washing should be included in estimating the possible trace element emissions from power plant stacks.

Second, it appears that some of the trace elements remain in the fly ash after the coal is burned in a power plant. The amount of a trace element remaining seems to be related, among other things, to the vapor pressure of the particular trace element being considered.

Precise and accurate trace element assays on coal are extremely difficult and require constant vigilance to avoid errors that can be introduced both in the laboratory and in the coal-handling procedures. In the future this program will extend the present studies to coals from different parts of the country, study additional trace elements in coal such as arsenic, selenium, beryllium, and others, and apply the developed techniques to more power plants.

Literature Cited

1. Deurbrouck, A. W., U.S. Bureau of Mines, private communication.
2. Schlesinger, M. D., Schultz, Hyman, "An Evaluation of Methods for Detecting Mercury in Some U.S. Coals," *Bur. Mines (U.S.) Rep. Invest.* (1972) **7609.**

3. Abernethy, R. F., Gibson, F. H., "Method for Determination of Fluorine in Coal," *Bur. Mines (U.S.) Rep. Invest.* (Dec. 1967) **7054.**
4. Diehl, R. C., Hattman, E. A., Schultz, H., Haren, R. J., "Fate of Trace Mercury in the Combustion of Coal," *Bur. Mines Tech. Prog. Rep.* (May 1972) **54.**
5. Roch, R. R., Gluskoter, H. J., Kennedy, E. J., "Mercury Content of Illinois Coals," *Environ. Geol. Notes* (1971) No. **43.**
6. Billings, C. E., Matson, W. R., "Mercury Emissions from Coal Combustion," *Science* (1972) **176,** 1232.
7. Bolton, N. E., *et al.,* "Trace Element Measurements at the Coal-Fired Allen Steam Plant," Progress Report, June 1971–January 1973, **ORNL-NSF-EP-43,** March 1973.
8. Kalb, William G., Baldeck, C., "The Determination of a Mercury Mass Balance at a Coal-Fired Power Plant," U.S. Environmental Protection Agency, Washington, D. C., December 30, 1973.

RECEIVED February 27, 1974. Reference to specific products is made to facilitate understanding and does not imply endorsement by the Bureau of Mines.

12

Total Mercury Mass Balance at a Coal-Fired Power Plant

G. WILLIAM KALB

TraDet Laboratories, Columbus, Ohio 43212

A series of mercury mass balances was obtained at a coal-fired power plant by comparing the volatile and particulate mercury in the stack gas stream to the mercury initially in the coal, corrected for the mercury adsorbed and retained by the various ashes. These data were used to determine the fate of the mercury in the combustion process and to check the accuracy of the volatile mercury sampling procedure (gold amalgamation). The bottom ash had the lowest mercury concentration of the ash samples collected, and the mercury concentration increased as one proceeded through the ash collection system from the initial mechanical ash to the electrostatic ash. The mercury recovered in the various ashes represented about 10% of the total mercury introduced in the raw coal.

Naturally occurring mercury volatilizes when fossil fuels are burned in power generating plants. The vapor pressure of mercury at the temperature ranges encountered in the duct work and stacks of these power plants is high enough that condensation does not occur at the mercury concentrations encountered. (The vapor pressure of mercury at 302°F is 18.3 mm Hg, which would represent a theoretical concentration of 26,000 ppm in air (1).) Based on the present coal consumption by the electric utilities in this country, approximately 350,000,000 tons/yr, and the average mercury concentration in coal of 0.1 ppm, approximately 35 tons of mercury are volatilized each year during coal combustion processes. With increased use of coal this figure will increase considerably. The purpose of this study was to obtain a series of mercury mass balances at a coal-fired power plant to determine the fate of mercury in the combustion process and to serve as a check on the accuracy of the volatile mercury sampling procedure.

Previous studies by the author (2, 3, 4, 5) and others (6, 7, 8) suggested that some of the mercury is retained or recovered in the various ashes before being released to the atmosphere. These studies also suggested that the amount of mercury released to the atmosphere is a function not only of the initial mercury concentration in the coal and the amount of coal consumed per unit of power produced but also of

1. The coal cleaning process
2. The furnace temperature
3. The type of bottom ash and flyash removal systems
4. The use of heat recovery systems in the facilities (*i.e.*, economizers, air preheaters, etc.)
5. The temperatures of the gas stream in the ash removal systems, duct work, and stack
6. The presence of any wet scrubbers in the sulfur dioxide removal systems.

Evaluating the quantitative effect of these factors on the volatile mercury concentration requires determining how much of the initial mercury found in the coal is: 1. not volatilized from the coal during combustion, 2. recovered in the various ash collection mechanisms by some adsorption phenomena, and 3. released to the atmosphere. With this information, a mercury mass balance can be calculated in which the amount of mercury consumed during the combustion process is compared with the amount in the stack gas and the various ashes. During this study, this was accomplished by comparing the stack gas concentration with the amount of mercury initially in the coal, corrected for the amounts recovered in the ashes. Differences between these two values would represent adsorption and/or desorption onto and off the walls of the ducts and stack and any significant contribution from the ambient air used in the combustion process.

The second objective of this study was to use the results of the mass balance to check the accuracy of the volatile mercury sampling procedure. The standard sampling procedure for mercury determination in stack gases (9) is based on collecting the volatile mercury in a series of impingers containing a liquid oxidizing solution, iodine monochloride. There are two major problems encountered with this procedure: high sulfur dioxide concentrations in stack gases reduce the iodine monochloride solution, greatly limiting the sampling time, and the analytical procedure for mercury determination in the iodine monochloride solution gives poor precision of replicated determinations because of a high response sensitivity that is subject to many procedural variations. A gold amalgamation sampling train for volatile mercury was developed (2, 3, 4, 5, 10, 11) to permit extended sampling in high sulfur dioxide environments such as smelters and power plants. The precision of this procedure has been

demonstrated by a series of simultaneously collected isokinetic samples
of stack gas streams (5). The mass balance (comparing the stack gas
concentration with the initial amount of mercury in the coal, corrected
for the amount of mercury recovered in the ash) was used here to check
the accuracy of this procedure.

Sampling Procedure

The isokinetic sampling procedure for volatile and particulate mercury used in this study consisted of collecting the volatile mercury by
an amalgamation reaction on gold foil chips supported by quartz wool
plugs in the impinger stems (amalgamators) of the standard EPA isokinetic sampling train. After the sample was collected, the mercury was
revolatilized by heating the amalgamator in an induction furnace and
was collected in an acidic permanganate solution that was analyzed by
flameless atomic absorption.

The results of the analyses of the solutions and the filter particulates
were used in conjunction with the stack gas flow rate to calculate the
mercury emission rate at the stack gas sampling point. A mercury in coal
consumption figure was obtained by monitoring the coal consumption
rate during the same test period and analyzing a representative sample
of the coal burned during the test period. During the same time interval,
ash samples were collected from the bottom of the furnace (bottom ash),
the initial and final mechanical ash hoppers, and the electrostatic precipitator. The mercury in ash concentration and estimates of the amounts
of ash collected were then used to correct the mercury in coal consumption
figure for comparison with the stack gas concentration. Both the mercury in the coal and ash were analyzed by a procedure reported by the
author (12) consisting of firing the sample in a quartz-enclosed graphite
crucible in an induction furnace and then collecting the volatilized mercury in an acidic permanganate solution which was analyzed by flameless
atomic absorption. Fourteen completed mass balances were obtained
at one coal-fired power plant using the above procedure. Two of the
tests were conducted simultaneously, permitting a check on both the accuracy and the precision of the stack gas sampling procedure.

Field Investigation

The generating unit investigated had a 93-megawatt output and
went on line in September 1949. The unit contained a pulverized coal,
corner-fired furnace with an Aerotec Corp. mechanical ash collection
system and a Research Cottrel electrostatic precipitator. At a 93-mega-
watt output, a steam flow of 810,000 lbs/hr resulted in a gas flow through
the electrostatic precipitators of 232,700 cu ft/min (CFM) in the north

duct and 220,300 CFM in the south duct. The measurements were made in the exiting ductwork at 358°F and 1.0–2.0 in. of water static pressure. Because of a nearby airport, the stack was relatively low with a venturi top which caused the positive pressure at the sampling point.

The initial sampling site (runs 1–12) was an existing 3-in. port on the west side of the south duct leading from the induced draft fan to the stack. This duct was one of two parallel ducts between the electrostatic precipitator and the stack. Three dual tests (13–18) were performed in the north side of the same duct where a 6-in. port had been installed.

Stack Gas Sampling. The gas samples from tests 1–12 were collected with a model 2343 RAC Staksamplr console and an EPA sample case modified to hold a maximum of eight impingers. Dual tests 13–18 used two model 2343 RAC Staksamplr consoles and an EPA dual train sampling box capable of holding five impingers on each side. Heated 5-ft glass probes were used with the sample cases. All of the trains used a probe connected to a 4¼-in. diameter filter which was enclosed in a heated compartment, then a series of impingers and/or amalgamators in an ice bath, followed by an impinger containing silica gel. The gold sampling train contained eight impingers: a Greenburg–Smith water scrubber, an empty impinger, four gold amalgamators, a potassium permanganate backup solution, and a silica gel impinger. The fourth amalgamator and the potassium permanganate backup solution were only used to measure the collection efficiency of the train and would not be used in a standard sampling train. Because of the smaller capacity of the dual boxes, only three amalgamators were used, and the backup solution was eliminated. One of the dual tests used a permanganate sampling train on one side which consisted of four potassium permanganate impingers and a silica gel impinger.

The standard sampling train illustrated in Figure 1 consisted of a Greenburg–Smith impinger containing 200 ml of distilled water, an empty impinger, and then a series of four gold amalgamators, each containing 25 g of 0.007-in. thick 1/16-in. square gold foil chips. The gold was heated in a refractory oven at 600–700°C for several hours between tests. Although a larger quantity of gold would show a higher individual amalgamator collection efficiency, the 25-g quantities were more sensitive to changes in the stack gas parameters, permitting observations of changes in the amalgamator collection efficiency caused by variations in moisture concentration, gas stream temperature, gold chip size, the presence of volatile organic compounds, etc. In actual practice, where the primary objective is an emission value, a larger quantity of gold would be used. The gold amalgamators were followed by 250 ml of either a 0.6% or 3.0% w/v potassium permanganate solution in 10% nitric acid and a silica gel impinger.

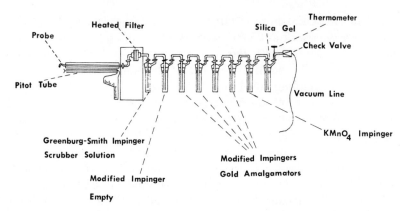

Figure 1.   Gold amalgamation isokinetic sampling train for mercury

The Greenburg–Smith water scrubber and the empty impinger in the first two positions of the train greatly decreased condensation in the amalgamators and cooled the gas stream. Without this scrubber the amalgamator must be dried before heating by passing a nitrogen stream through the amalgamator until the condensation disappears. In addition, without a scrubber, droplets condense in the train absorbing as much as 15% of the total mercury collected. This may be partially accounted for by analyzing the rinses of the impinger bases, but the droplets containing mercury on the inside of the amalgamator stem cannot be recovered and cause a significant decrease in the amount of mercury recovered from the train (3). Although the amalgamator shell rinses are analyzed when a scrubber is used, the amount of mercury recovered from these is normally insignificant (3).

After the train was assembled, the sample box was positioned at the port, the heater was turned on, the ice compartment was filled with ice and water, and a leak check was performed. The probe was then connected and inserted into the gas stream. Based upon isokinetic requirements, 0.187, 0.250, and 0.375-in. diameter probe tips were used at this site. Isokinetic runs were obtained over periods of 30–120 min with readings taken every 5 min.

After the sample was obtained, the probe and sample box were taken to the on-site mobile laboratory for the cleanup procedure. The mercury amalgamated on the gold was revolatilized by heating the amalgamator in an induction furnace and collected in 50 ml of a 3.0% potassium permanganate solution. The apparatus used for heating the amalgamators is shown schematically in Figure 2. The lower portion of the bubbler consists of an interchangeable 100 ml closed tube containing the potassium permanganate solution. The amalgamator was centered in the coil of the induction furnace (Leco model 521 with Variac control) and connected to a nitrogen supply and the bubbler by means of two female ball-joint adapters and clamps. The nitrogen flow was set at 0.5 l./min. Heating started at a Variac setting of 60% and was increased 5% each minute until the gold was glowing. This avoided a large spike of mercury into the bubbler and ensured complete heating of the gold.

After heating each amalgamator, the sample tube was detached, the drops of permanganate clinging to the bubbler were rinsed into it, and then the contents of the tube were transferred to a sample bottle. The amalgamators were heated in reverse order (first amalgamator in the train heated last) to minimize cross-contamination. After heating, the gold was poured into crucibles and placed in the refractory oven for several hours before reuse. The Tygon tubing connecting the amalgamator to the bubbler was replaced for each test to avoid contamination of succeeding tests by mercury adsorbed and desorbed from the Tygon.

The following samples were taken to account for all non-amalgamated mercury deposited in any part of the train ahead of the silica gel. Distilled water was used for all rinses unless otherwise noted.

1. Rinses of the probe and the glass parts of the filter assembly
2. The filter, previously weighed
3. Rinses of the inside portion of the first impinger and the right angle connection leading into it and the contents of the impinger shell containing the original 200 ml of distilled water and condensed moisture from the stack gas

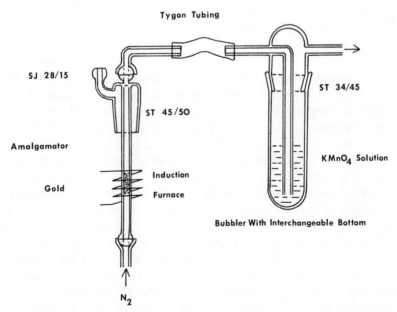

Figure 2. Apparatus for firing the amalgamators

4. Rinses of each empty impinger and the connector leading to it. This step was introduced when it was discovered that moisture condensed in an empty impinger often contained appreciable amounts of mercury, particularly if that impinger was ahead of the first amalgamator.

5. Rinses of each amalgamator case and its leading connector

6. The potassium permanganate impinger solutions. Any permanganate stain remaining in the impinger shell was removed with a few drops of 10% w/v hydroxylamine hydrochloride followed by a rinse with distilled water. These rinses were added to the permanganate in the sample bottle.

7. A 50 ml blank of the permanganate solution

A stock solution of 3.0% potassium permanganate in 10% nitric acid (prepared fresh daily) was used to stabilize the mercury collected in the rinses from the sampling train. The stabilized solutions were then returned to TraDet's Columbus, Ohio laboratory for analysis.

Coal Sampling. Representative coal samples were obtained during each run by taking approximately equal volume samples from each of the four feeders below the coal bins every 5 min during the test period. These samples were composited in a 10 gal milk can during the test. When the test was completed this sample was put through a large riffle splitter to obtain a representative 16-oz sample of the composite. The sample was then sealed and returned to the laboratory for drying, further splitting, pulverizing, and analysis.

The feeder revolutions were recorded every 5 min by counters mounted on each feeder. They had been previously calibrated to deliver 50 lbs/revolution. These values were then used to determine the total coal consumption during the test period.

The same individual collecting the coal samples and maintaining the feed rates also recorded the air flow–steam flow rates in lbs/hr every 5 min. These values were used in conjunction with previously obtained flow data to approximate the stack gas flow rate during the test period.

Ash Sampling. Four ash samples were to be obtained during each test: bottom ash, initial mechanical ash, final mechanical ash, and electrostatic ash. The bottom ash was deposited mechanically below the grates in the furnace, and the remaining ash samples were deposited in banks of ash hoppers below the duct work or electrostatic precipitators. These banks of hoppers were emptied through vacuum lines operated on a cycle, emptying only one hopper at a time. To obtain a representative sample of the ash during each test period, the three hoppers to be sampled had to be emptied just before the run. Because it was impossible to do this, normally only three of the four ash samples were obtained during any one test. With the type of ash collection systems used, it was impossible to obtain quantitative values on the amounts of various ashes accumulated per test period.

The bottom ash was sampled by emptying one of the four grates below the furnace at the start of the run. At the completion of the run this grate was re-emptied onto the floor. This sample was allowed to cool, and then a 10-gal milk can was filled with the ash sample. This sample was then split with a riffle splitter to obtain a 16-oz representative aliquot.

The mechanical and electrostatic ash samples were obtained by emptying the specified vacuum-operated hoppers just prior to the start of the run. When the test was completed, a cap plug was unscrewed from the bottom of the hopper permitting the hot ash to flow out. A 10-gal sample of each of these ashes was collected and split, maintaining a 16-oz sample of each ash. With the amount of ash involved, it was impossible to obtain a representative sample of the total amount of ash collected in the hopper during the test period.

Laboratory Analytical Procedure

The stack gas samples (including the particulate and volatile mercury), the ash samples, and the coal samples were analyzed by flameless atomic absorption using an LDC Mercury Monitor. The volatile mercury collected in the gold amalgamation sampling train was received in the laboratory in solutions stabilized with acidic permanganate. Depending upon the mercury concentrations, they were either diluted or analyzed directly by the stepwise reduction of the excess permanganate and the oxidized mercury. The reduced mercury was then entrained by aeration and carried through the absorption cell by the direct aeration technique. The filter particulates were digested by a nitric acid procedure and analyzed by the same direct aeration technique. The mercury in the coal and ash samples was volatilized by firing the samples in an induction furnace. The revolatilized mercury was collected in an acidic permanganate oxidizing solution. The mercury in these permanganate solutions was measured by the same procedure as was used for the volatile and filter–particulate samples.

Volatile Mercury Analysis. Upon receiving the gold amalgamation sampling train samples in the laboratory, the volume of the probe and filter holder washings was measured and recorded along with the total volume of each of the other solutions, with the exception of those obtained from heating the amalgamators. These were diluted to 100 ml in a volumetric flask.

Before each aliquot of sample solution was removed, the container was shaken thoroughly until all solids were evenly dispersed in the solution. A sample was quickly pipetted from the container and placed in an interchangeable sample tube where it was diluted to approximately 50 ml with distilled water. Three ml of hydroxylamine hydrochloride solution (10% w/v) were added, and the tube was swirled until the permanganate color disappeared. The tube was then attached to the bubbler and the solution was reduced with 3 ml of stannous chloride solution (20% SnCl$_2$ in 50% HCl). The mercury was then volatilized by aerating the solution with approximately 1.4 l./min of nitrogen. The revolatilized mercury was carried by the nitrogen stream through the mercury monitor. All samples were analyzed in duplicate using the equipment illustrated in Figure 3. A standard curve was prepared in duplicate from 50 ml aliquots containing 0.05, 0.10, 0.25, 0.50, and 0.75 μg of mercury. These standards were prepared from a standard 50 ppb mercury solution which was prepared fresh daily from a 1000 ppb standard prepared fresh weekly. The sample concentrations were determined from the standard curve, and the amount of mercury in each sample was calculated from the dilution factor and the size of the aliquot.

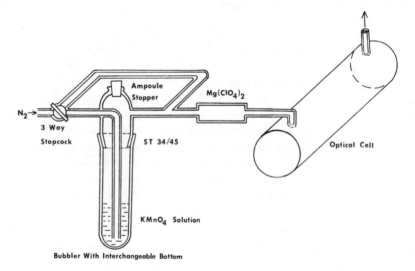

Figure 3.   Apparatus for analyzing the mercury in the permanganate-stabilized solutions

Filter Particulate Analysis. In the laboratory, the filters were immediately placed in a desiccator for 24 hrs and then weighed. The particulate concentration was calculated from this weight and the previously determined tare weight of the filter. The filters were analyzed for mercury by a procedure previously reported by the author (*11*) consisting of boiling the filters in concentrated nitric acid. After being cooled, aliquots of these solutions were pipetted into the interchangeable sample tubes and analyzed by the direct aeration technique.

Coal and Ash Analysis. The coal and ash samples were separated into 20–40 g representative samples in the laboratory by a small riffle splitter. These samples were then pulverized with a mortar and pestle. The moisture content was determined by weighing the sample, drying it at 103°C, and then reweighing it. These samples were kept in a desiccator and were subsequently used for the mercury analysis.

The mercury-in-coal analytical method used in this study was previously reported by the author (*12*). Approximately ½g-sample of the dried–pulverized coal was weighed on an analytical balance in quartz crucibles which had been heated in an induction furnace to remove any traces of mercury adsorbed on the crucible surface. The crucible was then placed in a quartz-enclosed graphite crucible which was placed in a Leco model 521 induction furnace. An air stream of 1.5 l./min was maintained through the quartz firing chamber which was connected to a bubbler similar to Figure 2. The interchangeable sample tube contained 50 ml of a 1.5% (w/v) potassium permanganate solution in 10% (v/v) nitric acid. The sample was initially fired at a 60% Variac setting in the furnace for 2 min and was then fired an additional minute at a 75% Variac setting. If the firing is commenced at the higher temperature (75% Variac setting) flashing will occur which may block the connection between the firing chamber and the bubbler or result in the coal sample being carried into the bubbler before firing is complete.

After the firing was completed, the permanganate solution was analyzed by the same procedure used for the volatile mercury samples. Mercury in coal samples from the National Bureau of Standards and the United States Bureau of Mines were analyzed simultaneously as controls. At least duplicate analyses were performed on all samples. Empty crucibles were fired into acidic permanganate solutions by an identical procedure to obtain solutions for blank determinations.

Mercury recovered in the ash probably represents mercury in the siliceous portion of the ash and mercury adsorbed onto the surface of the ash particles as the gas stream cools. Preliminary investigations showed that the coal analysis method does not account for all the mercury in the ash. This is a result of the significantly higher silicate concentration in the ash. In addition, the higher temperature ashes have a lower mercury concentration requiring a larger sample size for analysis. Because of these properties of the ash, the coal analysis method was modified to revolatilize the mercury quantitatively in the ash.

The method used in this study consisted of firing ½g-samples of the ash for 20 min at a 75% Variac setting in the Leco induction furnace. This procedure was repeated with a second ½g-sample without removing the permangate solution. Thus, each permanganate solution contained the mercury from a 1g-sample of ash. A stepwise increase in the firing temperature was not required with the ash samples because they do not flash.

The permanganate solutions obtained from both the coal and ash samples were analyzed by the direct aeration flameless atomic absorption procedure. The concentration in each sample was calculated from the measured mercury value, the blank concentration in the permanganate solutions, and the weight of the samples.

Results and Discussion

The results of this study were divided into three areas: stack gas analysis, solids sampling and analysis, and the mercury mass balance.

Stack Gas Analysis. The measured concentrations of mercury in the stack gas at the sampling point, including both the filter particulates and volatile mercury, are presented in Table I. Tests 15 and 18 were contaminated and should not be considered. Data from the fourth amalgamator and the potassium permanganate backup solution showed that the standard sampling train had a collection efficiency in excess of 98% of the mercury entering the probe.

Table I.   Mercury Concentrations in the Stack Gas

Run	Date (1973)	Total µg Hg	DSCF[a]	µg Hg/ DSCF[a]	Train Configuration[b]
1	7/16		9.910		H–E–A–A–A–A–K
2	7/16	1.896	9.399	0.202	H–E–A–A–A–A–K
3	7/17	2.271	16.787	0.135	H–E–A–A–A–A–K
4	7/17	2.056	16.759	0.123	H–E–A–A–A–A–K
5	7/17	6.428	40.437	0.159	H–E–A–A–A–A–K
6	7/17	3.852	32.528	0.118	H–E–A–A–A–A–K
7	7/18	8.226	72.841	0.113	H–E–A–A–A–A–K
8	7/18	4.690	25.078	0.187	H–E–A–A–A–A–K
9	7/18	7.818	38.463	0.203	H–E–A–A–A–A–K
10	7/19	8.622	46.982	0.184	H–E–A–A–A–A–K
11	7/19	4.076	24.673	0.165	H–E–A–A–A–A–K
12	7/19	3.444	24.426	0.141	H–A–A–A–A–K
13	7/20	8.151	52.272	0.156	H–A–A–A
14	7/20	7.292	50.485	0.144	H–A–A–A
15	7/20	25.262	53.164	0.475[c]	H–K–K–K
16	7/20	4.164	48.396	0.086	H–A–A–A
17	7/20	3.377	30.811	0.110	H–A–·A–A
18	7/20	24.967	32.373	0.771[c]	H–A–A–A

[a] DSCF = dry standard cu ft
[b] H = 200 ml distilled water
　E = empty impinger
　A = gold chips
　K = KMnO₄ solution
[c] Mercury contamination from probe tip

Table II. Coal Consumption During Test Periods

Run	Date (1973)	Coal Consumed During Test Period (lbs)	Moisture Content 103°C (%)	Dry Coal Consumed/ Test Period (lbs)	Dry Coal Consumed/ Test Period ($g \times 10^{-6}$)	Dry Coal Consumed/ Hr ($g \times 10^{-6}$)
1	7/16	32,900	4.3	31,500	14.30	28.60
2	7/16	31,650	4.3	30,300	13.76	27.52
3	7/17	34,800	3.6	33,500	15.21	30.42
4	7/17	33,250	5.1	31,600	14.35	28.70
5	7/17	85,050	4.5	81,200	36.86	29.49
6	7/17	67,650	5.3	64,100	29.10	29.10
7	7/18	82,700	5.2	78,400	35.59	23.49
8	7/18	69,700	3.2	67,500	30.64	30.64
9	7/18	52,350	3.0	50,800	23.06	30.67
10	7/19	131,500	3.3	127,200	57.75	28.88
11	7/19	64,900	4.6	61,900	28.10	28.10
12	7/19	63,600	5.0	60,400	27.42	27.42
13–14	7/20	68,200	5.3	64,600	29.33	29.33
15–16	7/20	65,250	4.5	62,300	28.28	28.28
17–18	7/20	64,300	4.1	61,700	28.01	28.01

The Greenburg–Smith water scrubber and the empty impinger behind it were added to the train to decrease condensation in the amalgamator. Previous studies at both an oil-fired power plant (3) and a smelter (11) resulted in the wet scrubber retaining 10–20% of the total mercury collected in the train. This study resulted in an average mercury retention rate of 46.7% in the wet scrubber. During a recent incinerator study by the author (5), the mercury retention rate in the wet scrubber averaged 67.9%. From the available data the increased retention rate in the scrubber appears to be a direct function of the moisture content of the stack. As the moisture content of the stack increased, the collection efficiency of the wet scrubber increased. The average moisture contents observed during the four previously discussed studies were: smelter, 0.05%; oil-fired power plant, low; this study, 5–8%; and incinerator, 13%. This relationship would be significant at power plants maintaining a wet scrubber for sulfur dioxide control.

Solids Sampling and Analysis. The coal consumption data is compiled in Table II. The ash, BTU, and sulfur analyses on the coal consumed during the 1-wk test period is presented in Table III. The data from the mercury in coal determinations are given in Table IV. Based on the authors experience with mercury in coal studies, the resulting concentrations are average for eastern United States coals. The author has found the spread in the individual coal determinations to be common when analyzing 0.5-g samples. The data in Tables III and IV show a relation-

Table III.   Proximate Analysis of Coal Consumed During Test Period

	Ash (%)	BTU/lb	Sulfur (%)
July 16–17			
as received	9.35	12,744	0.72
dried	10.04	13,684	0.77
July 18–20			
as received	10.64	12,856	0.94
dried	11.21	13,540	0.99

ship between the sulfur and the mercury in the coal. The composite coal sample obtained July 16–17, 1973 had a 0.77% sulfur content and an average mercury concentration of 0.074 ppm (μg/g), compared with 0.99% sulfur content and a 0.093 ppm mercury concentration in the July 18–20, 1973 composite sample.

The data from the mercury in ash determinations are presented in Table V. Four types of ash samples were analyzed and reported: bottom, initial mechanical, final mechanical (multicones), and electrostatic ash. Because of the inability to empty an initial mechanical, a final mechanical, and an electrostatic ash hopper simultaneously prior to the start of a test, normally only three ash samples were collected during any one test.

The data show that the bottom ash has the lowest mercury concentration of the various ashes and that the mercury concentration increases as one proceeds from the initial mechanical ash to the electrostatic ash. The average mercury concentrations of the ashes were: bottom, 0.035 ppm Hg; initial mechanical, 0.074 ppm Hg; final mechanical, 0.081 ppm

Table IV.   Mercury in Coal Concentrations

Run	Date (1973)	Coal Concentration (μg/g)					Mercury Consumed (g)[a]		
		1	2	3	4	Average	/test Period	/hr	/day
1	7/16	0.130	0.118	0.148	0.136	0.133	1.90	3.80	91.2
2	7/16	0.040	0.040	0.045		0.042	0.58	1.16	27.8
3	7/17	0.058	0.051			0.054	0.82	1.64	39.4
4	7/17	0.057	0.081			0.066	0.95	1.90	45.6
5	7/17	0.071	0.064			0.067	2.47	2.00	48.0
6	7/17	0.088	0.078			0.084	2.44	2.44	58.6
7	7/18	0.094	0.093	0.112		0.100	3.56	2.35	56.4
8	7/18	0.108	0.102			0.105	3.22	3.22	77.3
9	7/18	0.080	0.122	0.079		0.094	2.17	2.89	69.4
10	7/19	0.120	0.141			0.130	7.51	3.75	90.0
11	7/19	0.088	0.110			0.099	2.78	2.78	66.7
12	7/19	0.093	0.100			0.097	2.66	2.66	63.8
13–14	7/20	0.100	0.109			0.104	3.05	3.05	73.2
15–16	7/20	0.059	0.042			0.050	1.41	1.41	33.8
17–18	7/20	0.063	0.060			0.062	1.74	1.74	41.8

[a] Calculated from coal consumption data and mercury in coal concentrations

Hg; and electrostatic, 0.159 ppm Hg. The coal and ash concentrations are plotted in Figure 4 for those tests where at least one ash sample was collected. During this study the electrostatic ash always had a higher mercury concentration than the original coal, and the bottom ash always had a mercury concentration less than that of the original coal.

The results support the theory that the volatilized mercury is adsorbed or absorbed by the ash particles as the gas stream cools. Since the mercury concentrations in the stack gases are considerably lower than the theoretical mercury vapor pressure at the observed temperatures,

Table V.   Mercury in Ash Concentrations

Date	Test	Ash	$\mu g\ Hg/g$ of Ash
7/18/73	8	bottom	0.022
	8	initial mechanical	0.144
			0.100
	8	electrostatic	0.170
			0.187
7/19/73	10	bottom	0.053
	11	initial mechanical	0.095
	11	final mechanical	0.135
	11	electrostatic	0.193
	12	bottom	0.093
	12	initial mechanical	0.015
			0.036
	12	final mechanical	0.038
	12	electrostatic	0.143
7/20/73	13–14	initial mechanical	0.073
	13–14	final mechanical	0.070
	14–14	electrostatic	0.130
	15–16	bottom	0.002
			0.005
	15–16	initial mechanical	0.054
	15–16	electrostatic	0.131

condensation should not occur. The volatile mercury would then be removed by the suspended ash particles by some form of physical (van der Waals') and/or chemisorption forces. The increasing mercury concentration in the finer ashes is enhanced by the increased retention time of the finer ashes in the gas stream as compared with the mechanical ash and by the larger surface area of the finer ashes as compared with their mass. Based on these two factors, it can be seen that an increased rate of mercury removal by the ash would not be required to obtain a higher mercury concentration in the fly ash. Although the lower gas temperatures at which the finer ashes are removed may increase the rate of mercury removal by the suspended ash, it would not be required to obtain higher concentrations in the finer ashes.

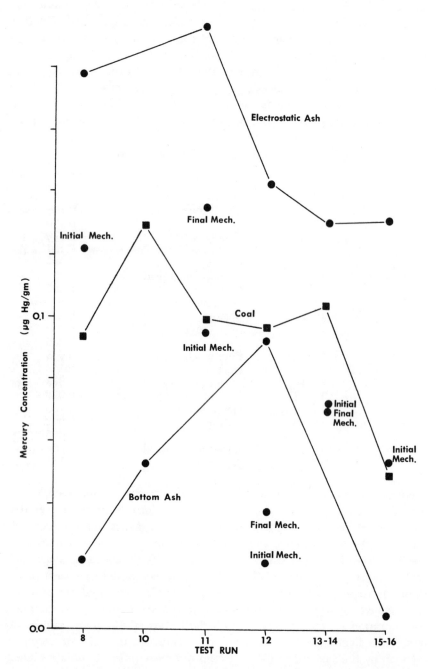

Figure 4.   Mercury concentrations in the coal and ash samples

The low concentration of mercury in the bottom ash results from one or a combination of the following: short retention time in the gas stream, the smaller surface area, and/or the higher temperatures at which it is removed from the system. The mercury concentrations in the filter particulates collected during the isokinetic sampling did not show any direct correlation with the ash studies because the mercury collected on the heated filter was partially revolatilized during sampling and collected in the remaining portions of the train.

Mercury Mass Balance. The mass balance consisted of comparing the amount of mercury in the stack gas stream with the amount in the coal minus that mercury recovered in the ash. In obtaining this balance it was assumed that the walls of the stack and the duct work did not affect the mercury concentration in the gas stream. The author felt that this was a correct assumption only if:

1. Sufficient time had elapsed since the start up of the unit so that the temperature of the walls was in equilibrium with the temperature of the gas stream

2. The unit was operating at a relatively constant load

3. The mercury adsorption mechanism was equilibrium controlled

Until these conditions have been achieved the rate of mercury adsorption on the walls will not equal the rate of desorption. The first two conditions were approximated in most of the tests reported. A second assumption in the mass balance was that the air utilized in the combustion process does not contribute a significant quantity of mercury to the system.

The two mercury values obtained for each run represent the grams of mercury per test run which pass the sampling point in the stack and which are in the coal that is not recovered in the various ash samples. If the two assumptions discussed in the preceding paragraph are correct, differences between the two reported values will represent the degree of representative sampling and the degree of accuracy in the analytical methodology.

The amount of mercury passing the sampling point in the stack was calculated from the measured stack gas concentration and the dry standard cu ft (DSCF) of gas passing the sampling point during the test period. The stack gas flow rate had been previously calibrated at a 93 megawatt output. At this load a steam flow of 810,000 lbs/hr resulted in a stack gas flow of 232,700 CFM in the north duct and 220,300 CFM in the south duct at 358°F and a 1.0–2.0 in. of water static pressure. Correcting for moisture and to standard conditions, this resulted in a combined gas flow of 285,788 dry standard cu ft/min (DSCFM) in both ducts at a steam flow of 810,000 lbs/hr. The DSCFM during the test period was calculated from the percent of the calibrated steam flow times the stack gas flow rate (285,788 DSCFM) at an 810,000 lbs/hr steam flow. The steam flow and air flow were maintained at the same lbs/hr rate. This flow approximation was used because the duct was not traversed during sampling.

The mercury in the coal not recovered in the various ash samples was determined by subtracting the amount of mercury recovered in the ash from the initial amount in the coal which was fed to the furnace during the test period. Because of the inability to measure the total amount of ash collected during the test period, the percent ash of coal (dried basis) was determined in the laboratory (*see* Table III). This percentage times the amount of coal consumed during each test period was used to determine the total amount of ash. The mechanical multicones had been tested at 40% effective and the electrostatic precipitators at 91% effective. But because of the unknown total amount of bottom ash these percentages could not be used to determine directly the amounts of the different ashes. The mercury-in-ash concentration used in each mass balance was 0.1 μg Hg/g of ash and was a weighted average based on the mercury concentrations measured in each ash and the available ash distribution data from the multicones and the electrostatic precipitator. Based on the mercury in the coal consumed during each test period and the mercury recovered in the ash, the average mercury retention by the ash was 13% of the total mercury consumed in the combustion process (range of 9–17%). Comparison of the total mercury-in-coal consumption and the mercury-in-stack gas emission values revealed that an average of 8% of the mercury present in the coal feed was not accounted for in the stack gas samples. The differences between the 13% ash retention rate of the mercury and the 8% unaccounted for from the coal and stack gas mercury values would represent experimental error, adsorption and desorption onto and off the walls of the stack and ducts, and any error in the ash distribution approximation.

The results of the mass balance are summarized in Table VI. The ratio of the mercury found in the gas stream to the amount of mercury released from the coal but not recovered in the ash is given in the third column. The average of this ratio is 1.19, with a standard deviation of 0.24. Within one standard deviation unit there is not a significant difference between the measured ratio and the expected ratio of 1.0. Therefore, based on 14 ratios at one plant, there is not a statistically significant difference between the total mercury found in the stack gas and the amount expected in the stack gas from the coal and ash determinations. Dual tests 13 and 14 showed close ratios of 0.88 and 0.95.

The two highest ratios, from tests 3 and 7, showed the greatest deviation from the expected norm of 1.0. Results from these tests showed a larger amount of mercury in the stack gas stream than was expected from the coal and ash analysis. These two tests were the first two tests on two different mornings and the kW output of the unit was either just being increased or had just been increased. During test 7 both the air flow and the stack gas temperatures were increasing during the test. The increasing temperature would result in the revolatilization of mercury adsorbed on the stack walls (nonequilibrium conditions). This would increase the ratio of the mercury found to the mercury expected. Kilowatt output and stack gas temperature were not routinely monitored between tests. As a result, the effect of these nonequilibrium conditions on the ratio could not be verified. Test 7 was the only test that showed significant changes in the stack gas temperature and air flow rate during the run.

Table VI.   Mercury Mass Balance

Run	Total Mercury Found in Stack Gas during Test Period (g)[a]	Total Mercury Expected in Stack Gas during Test Period (g)[b]	Ratio of Mercury Found to Mercury Expected
3	1.09	0.67	1.63
4	0.99	0.81	1.22
5	3.17	2.10	1.51
6	1.86	2.15	1.16
7	2.03	3.16	1.56
8	3.17	2.88	1.10
9	2.51	1.91	1.31
10	5.93	6.86	0.86
11	2.62	2.47	1.06
12	2.25	2.34	0.96
13	2.58	2.72	0.95
14	2.38	2.72	0.88
16	1.38	1.09	1.27
17	1.77	1.43	1.24

[a] As calculated from μg mercury/DSCF $\times$ DSCFM test period $\times$ minutes test period
[b] Grams mercury volatilized during test period and not collected on the ash

The data confirmed that in a coal-fired power plant there was not a significant amount of an unknown volatile organo–mercury compound bypassing the gold sampling train. Two assumptions were made at the start of the study—that equilibrium between the walls of the stack and the gas stream had been achieved and that the ambient air mercury concentration was not significant. The results have suggested that the first assumption is questionable. Although insufficient data are available to prove it, it is quite possible that the nonequilibrium conditions between the stack walls and the gas stream produces the spread in the ratio values. This results from unequal rates of mercury adsorption and desorption onto and off the walls. During future studies the kW output and the stack temperature should be recorded during the whole sampling period. At this time it is unknown if there is a cumulative buildup of mercury on the stack walls or if the concentration on the walls is controlled by a chemical equilibrium reaction. The data have suggested that when the stack temperature is increasing some mercury is released from the walls. The mercury concentration from ambient air should be checked in any future studies.

The mass balance could be improved by: obtaining gas flow rates by traversing the stack and determining the molecular weight of the gas stream, obtaining a method of measuring the quantities of the various ashes produced (isokinetic particulate sampling between each of the ash collection mechanisms), and calibrating the coal feeders immediately before the start of the tests. However, these improvements would considerably lengthen the required sampling time and would require a large field crew. In addition, more tests would result in better statistical results and would permit a more reliable measurement of the degree of accuracy. A complete inplant mass balance for mercury would require:

1. 24-hour monitoring of the kW output and stack gas temperature during the sampling period

2. Isokinetic sampling of the stack gas for the mercury and particulate concentrations

3. Isokinetic sampling of the ducts between each ash collection mechanism to determine the particulate loadings at each site

4. Sampling and analysis of samples from each ash collection mechanism

5. Sampling, analysis for mercury, and monitoring of the feedrate of the raw coal

6. Sampling, analysis for mercury, and flow monitoring of the influent and effluent of any wet scrubber systems

7. Monitoring of the mercury concentrations and flow rate of the ambient air used in the combustion process

Conclusions

A series of mass balances were obtained at a coal-fired power plant to determine the fate of mercury during the combustion of coal and to check the accuracy of the volatile mercury sampling procedure. The balance was obtained at the stack gas sampling point by comparing the volatilized and particulate mercury concentration in the gas stream with the amount of mercury in the coal fed to the furnace, corrected for the mercury recovered in the various ashes. Differences between these two values would represent:

1. Nonequilibrium adsorption and desorption onto and off the walls of the stack and duct

2. Any significant mercury concentration in the ambient air

3. The degree of representative sampling

4. The degree of accuracy of the analytical procedures

The average ratio of the mercury measured in the gas stream to the amount of mercury in the raw coal, corrected for the amount recovered in the ash samples, was 1.19 as compared with an expected ratio of 1.0. Based on the standard deviation, the measured ratio was not statistically different from the expected ratio. The data confirmed that in this plant there was not a significant amount of an unknown volatile organo–mercury compound bypassing the gold sampling train. The two highest ratios were correlated with increasing temperatures in the stack gas (increasing kW output) and would probably represent volatilization of mercury adsorbed on the walls of the stack and ducts at lower temperatures.

The results have shown that the fate of the mercury entering the system in the raw coal was controlled by the following mechanisms:

1. Some of the mercury in the coal was not volatilized during combustion, probably that mercury associated with the siliceous portion of the coal

2. A significant portion of the mercury ($\sim$10% of the volatilized mercury) was recovered by adsorption onto the suspended ash particles

3. Some of the mercury was removed from the system by adsorption onto the walls of the stack and ducts (it was not known if this was a cumulative or an equilibrium controlled mechanism

4. The major portion of the volatilized mercury was released to the atmosphere

The mercury adsorbed onto the ash may be differentiated from the non-volatilized mercury in the siliceous portion of the ash by firing the ash at different temperatures in the induction furnace. Because continuous changes in the kW output of the plant result in temperature variations in the stack gas, it was impossible to quantify the amount of mercury adsorbed onto the walls of the stack and ducts.

The results have suggested that mercury emissions would be decreased by: coal cleaning at preparation plants, improved heat recovery systems, longer retention times between the suspended ash and the gas stream, improved ash collection systems, and the use of wet scrubbers in the gas stream. Coal cleaning results in the removal of the higher specific gravity fractions of the coal, *i.e.*, pyrite, slate, etc. A significant percentage of the mercury in the coal is associated with this fraction and would thus be removed in the cleaning process before it reaches the power plant. Improved heat recovery systems in the plant (economizers, air preheaters, etc.) would lower the stack gas temperatures, permitting greater adsorption by the suspended ash particles. Longer retention times of the finer ashes in the gas stream would also increase the adsorption rate by the suspended ash. Improved ash collection systems would lower the amount of particulate mercury emitted to the atmosphere. Information obtained from the wet scrubber in the gold amalgamation sampling train has shown that significant quantities of volatilized mercury can be recovered in wet scrubbers. Most of these changes that have been shown to alter mercury emissions are presently being designed into new power plants to increase the plant efficiency, lower sulfur dioxide concentrations, decrease particulate emissions, etc. As a result, newer, more efficient power plants will probably show a lower percent mercury emission value.

Literature Cited

1. Jacobs, M. B., "The Analytical Toxicology of Industrial Poisons," p. 350, Interscience, New York, 1967.
2. Kalb, G. W., Baldeck, C., "The Development of the Gold Amalgamation Sampling and Analytical Procedure for Investigation of Mercury in Stack Gases," U.S. Environmental Protection Agency Contract No. **68-02-0341**, 1972.
3. Kalb, G. W., "The Determination of Total Mercury in the Emissions of an Oil Fired Power Plant," U.S. Environmental Protection Agency Contract No. **68-02-0225**, Task Order #17, 1973.
4. Kalb, G. W., Baldeck, C., "The Determination of a Mercury Mass Balance at Coal Fired Power Plant," U.S. Environmental Protection Agency Contract No. **68-02-0225**, Task Order #21, 1973.
5. Baldeck, C., Kalb, G. W., Kalb, M. A., "Comparison of Mercury Collection Techniques using Paired Runs at a Municipal Incinerator," U.S. Environmental Protection Agency Contract No. **68-02-0230**, Task Order #29, 1974.
6. Booth, M. R., Brown, J., unpublished data.
7. Diehl, R. C., Hattman, E. A., Schultz, H., Haren, R. J., "Fate of Trace Mercury in the Combustion of Coal," Bureau of Mines Managing Coal Wastes and Pollution Program, Technical Progress Report **54**, U.S. Dept. of the Interior, 1972.
8. Bolton, N. E., Carter, J. A., Emery, J. F., Feldman, C., Fulkerson, W., Hulett, L. D., Lyon, W. S., "Trace Element Mass Balance Around a Coal Fired Steam Plant," Oak Ridge National Laboratory, Oak Ridge, Tenn., NSF Interagency Agreement No. **AG-398**, 1973.
9. "National Emission Standards for Hazardous Air Pollutants: Asbestos, Beryllium and Mercury," Federal Register **38** (66), 1973.
10. Kalb, G. E., unpublished data.
11. Baldeck, C., Kalb, G. W., "The Determination of Mercury in Stack Gases of High SO_2 Content by the Gold Amalgamation Technique," U.S. Environmental Protection Agency # **EPA-R2-73-153**, 1973.
12. Kalb, G. W., "The Determination of Mercury in Coal by Flameless Atomic Absorption," *Amer. Chem. Soc. Div. Fuel Chem., Preprint* (1972) **16** (3).

RECEIVED February 27, 1974. This study was supported by the United States Environmental Protection Agency under Contract Number 68-02-0225 (Task Order No. 21).

Trace Element Mass Balance Around a Coal-Fired Steam Plant

N. E. BOLTON, J. A. CARTER, J. F. EMERY, C. FELDMAN,
W. FULKERSON, L. D. HULETT, and W. S. LYON

Oak Ridge National Laboratory, P.O. Box X, Oak Ridge, Tenn. 37830

Mass balance measurements for 41 elements have been made around the Thomas A. Allen Steam Plant in Memphis, Tenn. For one of the three independent cyclone boilers at the plant, the concentration and flow rates of each element were determined for coal, slag tank effluent, fly ash in the precipitator inlet and outlet (collected isokinetically), and fly ash in the stack gases (collected isokinetically). Measurements by neutron activation analysis, spark source mass spectroscopy (with isotope dilution for some elements), and atomic adsorption spectroscopy yielded an approximate balance (closure to within 30% or less) for many elements. Exceptions were those elements such as mercury, which form volatile compounds. For most elements in the fly ash, the newly installed electrostatic precipitator was extremely efficient.

Toxic elements are present in trace quantities in coal and other fossil fuels. Since enormous quantities of these fuels are consumed each year, appreciable quantities of the associated, potentially harmful toxic elements are produced. For example, if 600 million tons of coal are burned each year in the U.S. with average concentrations (ppm) of: Hg–0.10, Pb–20, Cd–0.4, As–5, Se–5, Sb–4, V–25, Zn–200, Ni–100, Cr–20, and Be–2, the corresponding tonnages of the elements released are: Hg–60, Pb–12,000, Cd–240, As–3000, Se–3000, Sb–2400, V–15,000, Zn–120,000, No–60,000, Cr–12,000, and Be–1200. (The concentrations are representative of values measured for coal burned at the Allen Steam Plant.)

An appreciable fraction (62%) of the coal consumed is burned at central power stations, so it is important to know the fate of potentially hazardous trace elements at such plants. The purpose of this work is to determine what happens to trace elements in coal which is used to generate electricity at a large central power station. The study involves two complementary activities: a mass balance for trace elements through the plant as obtained by in-plant sampling and measurements of the elements in the surrounding environment to estimate the effect of emissions on the concentration of toxic elements in air, soil, plant life, and in the water, sediment, and biota of the stream receiving the ash pond runoff. This paper deals with the in-plant portion of the work, which is a collaborative effort between ORNL and TVA.

This study was made at the Thomas A. Allen Steam Plant in Memphis, Tennessee, which has an 870 MW(e) peak capacity from three similar cyclone fed boilers. The plant is part of the TVA power system and was chosen because the Number 2 unit was being renovated. A new Lodge Cottrell electrostatic precipitator was being added so the TVA Power Production Division test sampling crew were available to help sample during compliance testing of the precipitator.

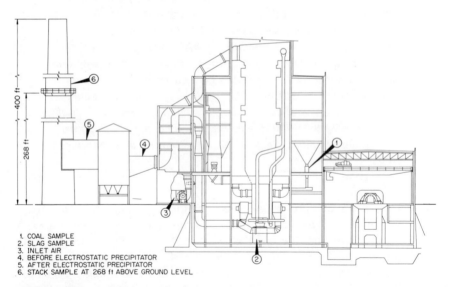

1. COAL SAMPLE
2. SLAG SAMPLE
3. INLET AIR
4. BEFORE ELECTROSTATIC PRECIPITATOR
5. AFTER ELECTROSTATIC PRECIPITATOR
6. STACK SAMPLE AT 268 ft ABOVE GROUND LEVEL

Figure 1.   Schematic of Number 2 unit, Allen Steam Plant, Memphis, Tenn.

Sampling and Methods of Analysis

Figure 1 shows the sampling points of the Number 2 unit. Samples taken at locations 1 and 2 were composite samples of the coal entering the boiler and of the slag material leaving the boiler, respectively. At location 3 the inlet air being supplied to the boiler was sampled. At loca-

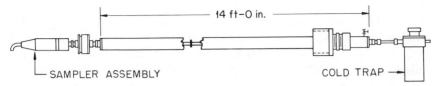

Figure 2.   Schematic of sampling probe used for mass balance study

tions 4 and 5 a series of samples was taken isokinetically at various locations in the ducts before and after the electrostatic precipitator, respectively. At location 6 a series of samples was taken isokinetically in the
stack approximately 82 m above ground level.

The large size of the ducts being sampled required specially fabricated sampling probes and special probes equipped with forward-reverse
pitot tubes for determining isokinetic sampling rates. Figure 2 shows
the construction of the sampling probe. The sample is drawn through an
alundum thimble at a predetermined isokinetic sampling rate. Behind
this thimble is a Gelman fiberglass filter paper holder which collects particles as small as 0.1 μ. This is essentially the standard ASTM method for
sampling gases for particulates (*1*). These filters are followed by the
cold trap which brings the flue gas through the dew point quickly and
collects all materials which will condense in a dry trap. Because of the
very high moisture concentration in the flue gas, a dropout glass jar was
added following the cold trap to prevent loss of the condensate. The
cold trap and condensate dropout jar were used to trap mercury and
other condensable vapors from the flue gas sample.

Two probes were fabricated for sampling the precipitator inlet, one
for the precipitator outlet, and one for the stack. By limiting the number
of test points at each plane in the precipitator outlet, the average sampling
time required to complete a test was about 280–300 min. The types
and numbers of samples collected for each complete run are shown in
Table I. Twenty-four distinct samples were taken for each complete
test run.

Table I.   Types and Numbers of Samples

Composite coal sample—1
Composite slag tank sample—1
Precipitator inlet—6 thimbles, 2 cold traps, 2 glass papers
Precipitator outlet—4 thimbles, 1 cold trap, 1 filter paper
Stack sample—2 thimbles, 1 trap, 1 glass paper
Inlet air—1 thimble and 1 glass paper

Although the plant is designed to operate at 290 MW per unit, a
240 MW load was chosen for these tests because it was felt that this load
could be maintained without interruption during the required 5-hr
sampling time. The coal consumption at this power level is 82.5 tons

per hr on a dry weight basis. The coal and slag was sampled by compositing samples obtained periodically during the test. The total particulate flow rate was calculated from the weight of fly ash material collected in the ASTM filter system, the total air volume passed through the thimble, and the velocity of air passing through the system. In all, four runs were made—one reference test and three for mass balance (runs 5, 7, and 9). A gas velocity traverse was made in the precipitator inlet and outlet ducts just prior to each mass balance run to determine isokinetic sampling rates for each sampling position.

The reference test was performed using the standard ASTM method for determining grain loadings to electrostatic precipitators. This test was used to verify the adequacy of the number of samples secured for mass balance calculations. Comparison of grain loading calculations using TVA standard probes and ORNL fabricated probes show that the mass balance samples are representative.

Analysis of the samples for elemental constituents was performed using instrumented neutron activation analysis (NAA) and spark source mass spectrometry (SSMS) (2). In addition, the many mercury determinations were made by flameless atomic absorption (AA).

The NAA technique involved irradiating each dry homogenized sample (0.01–0.2 g) in a sealed plastic vial. This vial was placed in a "rabbit" together with gold and manganese flux monitors and irradiated in the Oak Ridge Research Reactor for a period varying from a few seconds (for short lived radioactive products) to 20 min. After irradiation, the samples were removed and counted at various set decay times using a germanium–lithium detector and a nuclear data PDP-15 analyzer computer system. Using programs developed at ORNL, these counting data were processed, x-ray peaks identified, absolute activities calculated, and from the flux measurement and known nuclear parameters, the $\mu g/g$ of each element found was calculated. Results in all cases have an uncertainty of 5–10%. The entire process is nondestructive because no chemical treatment is performed, so there is only a minimal chance of sample contamination or loss.

The flameless atomic absorption method has a reproducibility of about 2% or better for homogeneous specimens. Checks (3) between AA and NAA (with radiochemical separation after irradiation) and isotope dilution spark source mass spectroscopy on thoroughly homogenized tuna fish and Bureau of Mines round-robin coal specimens indicate good agreement between the methods. (0.425 ± 0.9%, 0.45 ± 3.5%, and 0.45 ± 4.4% for tuna by AA, NAA, and SSMS, respectively, and 1.004 is the average ratio of NAA to AA results for five coal samples.) The similar results indicate that the technique used in sample preparation for AA did not result in mercury losses since the NAA method is not subject to losses of this type.

Spark source mass spectrometry (SSMS) is also a multielement technique. Conventionally the data obtained are semiquantitative, and the results have an uncertainty of ±50% or less. If the stable isotope dilution technique is performed, the SSMS can be ±3%. This latter technique was used for lead, cadmium, and zinc as noted in the results tabulations. NAA and SSMS complement each other quite well, and those elements for which one technique has poor sensitivity can usually be measured by the other.

Mass Balance Results

A mass balance for the various elements was calculated using the following equations:

$$Q_c(A) = C_c(A) \times (g \text{ coal/min}) \tag{1}$$

$$Q_{p.i.}(A) = C_{p.i.}(A) \times (g \text{ fly ash/min}) \tag{2}$$

$$Q_{s.t.}(A) = C_{s.t.}(A) \times (g \text{ ash in coal/min} - g \text{ fly} \atop \text{ash to precipitator/min}) \tag{3}$$

for balance:

$$Q_c(A) = Q_{P.I.}(A) + Q_{s.t.}(A) \tag{4}$$

$$\text{percent imbalance} = \frac{Q_{P.I.} + Q_{S.T.} - Q_c}{Q_c} \times 100 \tag{5}$$

where $Q_c(A)$, $Q_{P.I.}(A)$, and $Q_{S.T.}(A)$ are the flow rates of element A in g/min associated with the coal, precipitator inlet fly ash, and slag tank solids, respectively, and $C_c(A)$, $C_{P.I.}(A)$, and $C_{S.T.}(A)$ are the corresponding concentrations of element A in the coal, the fly ash collected in the precipitator inlet, and the slag tank solids. The flow of trace elements into the plant with suspended particulates in inlet air was negligible. We were unable to measure the total solids flow from the slag tank because of the nature of this discharge. (Every 4 hrs the slag tank residue is washed out to the ash pond with 2–4 hundred thousand gallons of water.) For this reason, we estimated the slag tank discharge as the difference between ash flow rate in the coal and the total fly ash flow rate. Presuming that this assumption is valid, that the sampling was complete and representative, and that the analyses are correct, the condition for balance is given by equation 4. To test this we have calculated a percent imbalance from experimental results by equation 5. Also, the precipitator efficiency for an element was calculated by:

$$\text{Precipitator efficiency} = \frac{Q_{\text{P.I.}}(\text{A}) - Q_{\text{P.O.}}(\text{A})}{Q_{\text{P.I.}}(\text{A})} \times 100 \qquad (6)$$

The results of the mass balance calculations for eight major elements and 22 minor elements for run 9 are given in Tables II and III, together with the corresponding concentrations in the coal, precipitator inlet and outlet fly ash, and in the slag tank solids. Complete tabulation of results for all three runs and some data for 57 elements is given in a project progress report (4).

In general, agreement between the two analytical methods is reasonable. There is a consistent negative imbalance, averaging −26% and −16% for NAA and SSMS results, respectively, for the major elements and −1% and −18% for the minor elements. We excluded the results for mercury and arsenic in the averages for minor elements. In view of the necessary assumptions and the difficulty of obtaining truly representative samples, the balance is satisfactory for most elements. Notable exceptions are elements which can be present in a gaseous form. One may be arsenic (Table III), and another is mercury which is discussed

Table II. Elemental Concentrations and Mass Balance

		Concentration (ppm unless otherwise indicated)			
Element	Method	Coal	S.T.[a]	P.I.[a]	P.O.[a]
Al	NAA	1.06%	6.6%	6.9%	3.5%
	SSMS	1%	5%	15%	10%
Ca	NAA	0.38%	2.7%	1.4%	0.49%
	SSMS	0.5%	3%	3%	1%
Fe	NAA	1.3%	10.1%	9.3%	23.5%
	SSMS	2%	10%	10%	10%
K	NAA	0.22%	0.95%	1.65%	1.28%
	SSMS	0.06%	0.5%	0.7%	0.2%
Mg	NAA	0.17%	0.41%	0.55%	0.88%
	SSMS	0.15%	0.7%	0.7%	0.4%
Mn	NAA	54	418	323	550
	SSMS	100	1000	700	500
Na	NAA	0.069%	0.32%	0.7%	0.28%
	SSMS	0.03%	0.2%	0.3%	0.2%
S	NAA	5.1%			10.5%
Si	SSMS	5%	30%	30%	10%
Ti	NAA	710	3000	3700	2500
	SSMS	700	2000	5000	1000

[a] S.T., P.I., and P.O. are abbreviations for slag tank solids, precipitator inlet, and precipitator outlet, respectively.

below. One reason for the consistent negative imbalance could be that fly ash samples were taken under steady state conditions. Two operations were not investigated, and these might account for this imbalance. The air heaters are cleaned pneumatically once per eight-hour shift, and soot is blown from the boiler tubes about two times per shift. If this material were measured it would increase the average fly ash flow rate ($Q_{P.I.}$). It is not known whether or not these operations can account for a significant percentage of the trace elements. Future in-plant sampling will include these two operations.

As in the case of the slag tank, there was no way to measure quantitatively the precipitator residue flow rate. These residues are slurried with water and flushed continuously into the ash pond. However, for all of the elements except selenium, the precipitator was extremely efficient (>95%) as calculated from the inlet and outlet fly ash concentrations using Equation 6. The reason that selenium fails to be scavenged effectively is not known and certainly warrants investigation. One possibility is that part of the selenium is in a volatile state but is readily adsorbed on particulates trapped by the alundum thimbles.

Mercury has been determined on virtually every sample (the filters, cold trap and slag tank water, and residue). We were unable, however,

Results for Major Elements Measured for Run 9

Mass Flow (g/min)

Coal	S.T.[a]	P.I.[a]	Imbalance (%)	P.O.[a]	Precipitator Efficiency (%)
1.3×10^4	7.2×10^3	3.4×10^3	-18	68	98
1.3×10^4	5.5×10^3	7.3×10^3	-1	190	97
0.47×10^4	3.0×10^3	6.8×10^2	-22	9.5	99
0.6×10^4	3.3×10^3	1.5×10^3	-20	19	99
1.6×10^4	1.1×10^4	4.5×10^3	-3.1	460	90
2.5×10^4	1×10^4	5×10^3	-40	190	96
0.27×10^4	1.0×10^3	8.0×10^2	-33	25	97
0.07×10^4	$5. \times 10^2$	3×10^2	$+14$	4	99
0.21×10^4	4.5×10^2	2.7×10^2	-66	17	94
0.18×10^4	8×10^2	3×10^2	-39	8	98
67	46	16	-7.5	1.1	94
130	110	34	11	1	97
860	350	340	-20	5.5	98
370	220	150	0	4	97
6.4×10^4				200	
6.3×10^4	3.3×10^4	1.5×10^4	-24	190	99
890	330	180	-43	4.9	97
880	220	240	-48	2	99

Table III.   Elemental Concentration and Mass Balance

Element	Method	Concentration (ppm)			
		Coal	S.T.[a]	P.I.[a]	P.O.[a]
As	NAA	3.8	0.5	46	50
	SSMS	5	2	40	20
Ba	NAA	79	600		
	SSMS	100	300	1700	100
Be	SSMS	<5	<10	17	<10
Cd	ID[b]	0.47	~3	5.8	
	SSMS	0.5	2	<10−20	7
Co	NAA	3.3	19	25	58
	SSMS	7	40	70	40
Cr	NAA	21	180	356	300
	SSMS	30	<200	70	40
Cs	NAA	1.5	8	21	4
Cu	SSMS	50	200	400	400
Eu	NAA	0.17	1.4	1.8	
	SSM	~1			
Hg	AA	0.063	0.09	0.043	
La	NAA	5.0	42	32	
	SSMS	~10			
Li	SSMS	25	200	300	200
Mo	NAA	20			
	SSMS	20	80	200	20
Ni	SSMS	≤100	500	500	1000
Pb	ID[b]	7.4	~4	149	
	SSMS	<20	3	250	100
Sb[d]	NAA	<1	<0.2	3.2	
	SSMS		8	7	10
Sc	NAA	3.2	22	25	10
Se	NAA	3.2	14	<32−48	760
	SSMS	6	20	20	200
Sn[e]	SSMS	20	200	20	20
Th	NAA	3	20	18	
Tl	SSMS	<2	2	40	30
U	NAA	1.67	14	17	7
V	NAA	21	125	200	63
	SSMS	30	100	350	100
Zn	ID[b]	94	~20	1500	
	SSMS	85	100	3000	900

[a] S.T., P.I., and P.O. stand for slag tank solids, precipitator inlet, and precipitator outlet, respectively.

[b] Data obtained by isotope dilution mass spectrometry for composite samples from runs 5 and 9.

[c] Precipitator efficiencies calculated on the basis of P.O. by SSMS for run 9 and P.I. by ID from the composite sample runs 5 and 9.

[d] Run 5 data.

[e] Run 7 data.

Results for Minor Elements Measured for Run 9

Mass Flows (g/min)

Coal	S.T.[a]	P.I.[a]	Imbalance (%)	P.O.[a]	Precipitator Efficiency (%)
4.7	0.05	2.2	−52	0.1	95
6.2	0.2	2	−64	0.04	98
99	66				
130	33	83	−11	0.2	99+
∨6.3	<1.1	0.83		>0.02	98
0.59	0.31	0.33	+8.5		
0.63	0.22	<0.5−1.0		0.014	≤96ᶜ
4.1	2.1	1.2	−19	0.11	91
9	4.4	3.4	−13	0.08	98
26	20	17	42	0.6	96
37	<22	3.4		0.08	98
1.9	0.88	1.02	0	0.008	99+
63	22	19	−35	0.8	96
0.21	0.15	0.09	14		
~1					
0.079	0.0099	0.0021	−85		
6.3	4.6	1.5	−3		
~10					
31	22	15	19	0.4	97
25					
25	8.8	9.7	−26	0.04	99+
≤130	55	24		2	92
9.25	0.42	8.60	−3		
<25	0.3	12		0.2	98ᶜ
<0.75	<0.02	0.2			
	0.8	0.5		0.02	96
4.0	2.4	1.2	−10	0.02	98
4.0	1.5	<1.5−2.3	<−25 to −5	1.4	7–39
7.5	2.2	1.0	−58	0.4	60
25	20	1.4	−14	0.04	97
3.7	2.2	0.87	−17		
<2.5	0.2	1.9			
2.1	1.5	0.83	11	0.014	98
26	14	9.7	−9	0.12	99
37	11	17	−24	0.2	99
117	2.1	86	−24		
110	11	150		1.7	98ᶜ

to find the bulk of the mercury that we knew was entering the system *via* the coal. From this we concluded that mercury is present in the stack gas as a vapor which we were unable to trap.

Table IV gives all of the values obtained for mercury in coal, which range from 0.057 to 0.198 ppm with most values in the range of 0.07 ppm. Our attempt at a mercury balance for runs 5 and 9 is shown in Table V. From these numbers it is clear that very little mercury ($\sim 12\%$) remains with the slag and fly ash particles. The cold trap was not effective in trapping mercury vapor ($\sim 11\%$). The results are in qualitative agreement with those of Billings and Matson (5), except that these authors were able to collect the mercury in the gas phase. Their data shows that most of the mercury is in the gas phase. This can also be implied from our results.

Table IV. Mercury in Coal as Determined by Atomic Absorption

Sample	Date	$Hg(\mu g/g)$		
2 ECS 24	26 Jan	0.057		
5 CS 24 AM	28 Jan	0.064	0.063	
5 CS 24 PM	28 Jan	0.069	0.058	
7E CS 26 AM	31 Jan	0.198		
7E CS 26 PM	31 Jan	0.169	0.148	0.136
9E CS 16 AM	1 Feb	0.076	0.060	
9E CS 16 PM	1 Feb	0.060	0.058	
10E CS 19 AM	2 Feb	0.068		
10E CS 19 PM	2 Feb	0.073		
11E CS	3 Feb	0.060		

Recently, we returned to the Allen plant and sampled the flue gas using a four impinger train with a pre-scrubber of sodium carbonate to remove the acid gases, followed by three impingers charged with iodine monochloride solution. Preliminary results show that mercury quantities detected were of the expected magnitude based on mercury concentrations in the coal which we had measured previously. This technique will be used for the mercury balance at the next in-plant sampling.

Fly Ash Particle Characterization

Figure 3 shows scanning electron photomicrographs of fly ash particles from the precipitator inlet and outlet and from the stack. The particles are predominantly spherical, and there is considerable agglomeration of small particles (submicron size) to large ones. Also, there appears to be a fuzzy material present which might be a sulfur compound. Preliminary evidence for this is scanning electron microscope fluorescence analyses of some of the larger particles deposited from the precipitator

Table V.   Mercury Balance

Material	Average Flow (g/day)	Hg (μg/g)	Hg Flow (g/day)
	Run 5		
Coal	1.8×10^9	0.064	115
Ash (slag)	1.43×10^8	0.07	10
Precipitator inlet	0.96×10^8	0.04	4
H₂O to ash pond	2.9×10^9	0.003	9
Gas (cold trap)	4.3×10^{10}	0.0003	13
	Run 9		
Coal	1.8×10^9	0.064	115
Ash (slag)	1.58×10^8	0.09	14
Precipitator inlet	$0.70 \times 10_8$	0.043	3
H₂O to ash pond	2.9×10^9	0.001	3
Gas (cold trap)	4.4×10^{10}	0.0003	13

inlet flue gas on the first stage of a Cassella cascade impactor. Figure 4 shows such an analysis. All of the fluorescence lines, except aluminum, can be attributed to the particles. Since the particles were collected on an aluminum foil, the aluminum peak is caused primarily by the foil. After ion etching by bombardment with argon ions, the sulfur peak decreased substantially, indicating that sulfur was present primarily on the surface of the particles. As one would expect, the preliminary evidence is that the fly ash particles are a complicated mixture of the elements.

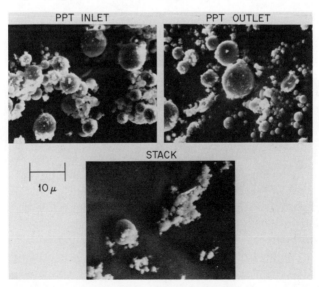

Figure 3.   Scanning electron photomicrographs of fly ash particulates collected on alundum thimbles used to sample the precipitator inlet and outlet and the stack

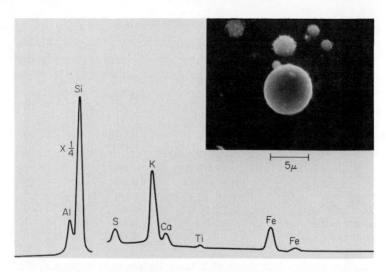

*Figure 4.   Qualitative fluorescence analysis of a 5 μ fly ash particle
trapped on the first stage of the Cassella cascade impactor used to
sample the precipitator inlet*

Work is still in progress on determining the particle size distribution
in the flue gases before and after the precipitator and in the stack. Also,
studies on fly ash composition as a function of particle size are in progress.

Conclusions

Trace element mass balance measurements around the Number 2
Unit of the coal-fired Allen Steam Plant in Memphis yielded a respectable
balance for many elements. However, the results showed a consistent
negative imbalance. This might be caused by the fact that soot blowing
and air heater cleaning operations were not taken into account in the
sampling. Because the flue gas sampling method was designed primarily
to collect particulates efficiently, good balances were not obtained for
elements forming volatile compounds. For example, more than 80% of
the mercury entering with the coal is emitted with the flue gas as a vapor.
The large imbalance for arsenic (-58%, Table III) indicates that a
substantial portion of this element is also in the vapor phase of the flue gas.

The electrostatic precipitator was very efficient ($\sim 98\%$) for most
trace elements based on analyses of the fly ash particulate specimens
collected from the precipitator inlet and outlet. An exception was selenium. Although a reasonable mass balance was obtained for this element (*see* NAA results, Table III), it was not removed efficiently by the
precipitator. This may indicate that a significant fraction of the material
is in the vapor phase in the flue gas and that it is being adsorbed in

passing through the alundum thimble filter used to sample the fly ash. Accounting more completely for the volatile trace elements such as mercury, selenium, and arsenic remains the most significant question still to be answered in future mass balance work.

Acknowledgments

The authors express their appreciation to the following personnel of the Tennessee Valley Authority without whose assistance this study could not have been accomplished: Joseph Greco, Chief Plant Engineering Branch, Division of Power Production; John H. Lytle, Plant Engineering Branch, who supervised the field test crew during the mass balance runs; and Lucy E. Scroggie, Supervisor of the Industrial Hygiene Laboratory, who coordinated TVA's contribution in this joint study for F. E. Gartrell, Director of Environmental Planning; and Carl J. Bledsoe, Superintendent at the Allen steam plant and his staff.

This work was supported by the National Science Foundation RANN Program under NSF Interagency Agreement No. AG-398 and AG-450 with the U.S. Atomic Energy Commission. Oak Ridge National Laboratory is operated by Union Carbide Corp. for the U.S. Atomic Energy Commission, Contract No. W-7405-eng-26.

Literature Cited

1. *Book ASTM Stand.*, "Standard Method for Sampling Stacks for Particulate Matter," ASTM Designation, **D-2928-71**, Jan. 8, 1971.
2. Carter, J. A., Sites, J. R., "Analysis of Radioactive Samples by Spark Source Mass Spectrometry," in "Trace Analysis by Mass Spectrometry," (A. H. Abern, Ed.), p. 347, Academic, New York, 1972.
3. Feldman, C., Carter, J. A., Bate, L. C., "Measuring Mercury," *Environment* (1972) **14**, (6), 48.
4. Bolton, N. E. *et al.*, "Trace Element Measurements at the Coal-Fired Allen Steam Plant, Progress Report, June 1971–January 1973," Oak Ridge National Laboratory report, **ORNL-NSF-EP-43**, March 1973.
5. Billings, C. E., Matson, W. R., "Mercury Emissions from Coal Combustion," *Science* (1972) **176**, 1232.

RECEIVED January 21, 1974

14

Major, Minor, and Trace Elements in the Liquid Product and Solid Residue from Catalytic Hydrogenation of Coals

P. H. GIVEN, R. N. MILLER, N. SUHR, and W. SPACKMAN

College of Earth and Mineral Sciences, Pennsylvania State University, University Park, Penn. 16802

Approximate contents of 14 minor and trace elements in oils produced from three coals by the catalytic hydrogenation process of Gulf Research and Development Co. were determined by emission spectroscopy. The results were compared with corresponding data for the original coals and the solid residues from the process. The contents of ash, sulfur, vanadium, lead, and copper are near or below the limits specified for an oil to be fired directly in a gas turbine while sodium and probably calcium are too high. Titanium appears to be somewhat enriched in the oils analyzed relative to other elements, suggesting its presence in organo-metallic complexes.

The ash content and trace element distribution in oils produced from coal are of concern for two different reasons—they bear on possible environmental hazards from the use of the oil and they determine the suitability of the oil for firing in gas turbines. Some trace element analyses of oils produced by the catalytic liquefaction of three coals are reported, together with analyses of the solid residues from the process. The data are neither comprehensive nor particularly accurate and are offered at this time because, in the absence of better information from other laboratories, they seem to be of some interest.

Procedure

The samples of oil and residue were produced in the course of the cooperative research project between workers at this university and at

188

Gulf Research and Development Co. Two of the coals were processed in the Gulf continuous flow reactor, fed at the rate of about 1.5 kg coal/hr for 15–18 hrs. The third coal was processed in a conventional batch autoclave run. In all three runs, the coal was processed at about 400°C and 3000 psi pressure of hydrogen using a proprietary catalyst. In the continuous runs, distillate from previous experiments was used as vehicle while in the autoclave experiment, partly hydrogenated phenanthrene was used. The vehicle-to-coal ratio was 2:1. In each case the reaction products were filtered on a steam-heated Buchner funnel.

The oils were distilled at 1 mm pressure until about one third remained in the pot. The undistilled residue, the solid residue from the reaction (catalyst pellets removed by hand), and the original coal were ashed in an oxygen plasma without external heating, using a low temperature asher from International Plasma Corp.

The ashes were analyzed by emission spectroscopy using a d.c. arc and a Jarrell–Ash Wadsworth spectrograph. A visual comparison of line intensities with standard rock samples established the approximate elemental concentrations. Based on experience, the true content of any element is most probably within the range 2 to 0.5× the stated value.

The coals were collected in the field by W. Spackman and his associates. Full petrographic, proximate, and ultimate analyses are available, but it does not seem appropriate to reproduce the full data here. The mineral matter contents reported below were determined by acid demineralization (1).

Results and Discussion

The results are shown in Tables I, II, and III. The major elements in coal and in the derived products are, in order of decreasing abundance in the materials studied, silicon, aluminum, calcium, iron, and magnesium. With the minor and trace elements, the detection limits vary with the ash content of each type of material; about 1–5 ppm for the coals and residues and 1–3 ppb for the oils. The elements thallium, bismuth, germanium, and gallium were sought but not detected.

Clearly the oil is a much cleaner fuel than the original coal, from the point of view either of the environmentalist or of the plant engineer concerned with fouling of steam and superheater tubes. The sulfur contents of the oils are 0.1–0.2%, which is acceptable, but the nitrogen contents are about 0.6%, which may cause undesired NO_x emissions. Some of the more toxic elements, (mercury, selenium, fluorine, and cadmium) have not yet been determined in oil. It is not clear what will be done with the solid residue: whether it will be disposed of as waste or whether its small carbon content, typically 20–50% depending on the

Table I.   Concentration of Minor and Trace Elements

	B	Co	V	Ni	Ti	Mo
Feed coal	12	25	12	12	75	tr.
Solid residue	25	12	25	5	150	25
Oil	0.15	0.12	0.06	0.15	5	0.08

[a] Mineral matter, dry basis, 12.9%. High Vol. B. Total S, 0.52 (S_{pyr} 0.13). Ash content of oil, 150 ppm.

Table II.   Concentration of Minor and Trace Elements from

	B	Co	V	Ni	Ti	Mo
Feed coal	25	12	50	25	400	12
Solid residue	150	40	150	80	1100	40
Oil	0.12	0.01	0.25	0.08	6	0.01

[a] Mineral matter, 25%. High Vol. A. Total S, 5.2% (Spyr 4.0). Ash content of oil 250 ppm.

Table III.   Concentration of Minor and Trace Elements

	B	Co	V	Ni	Ti	Mo
Feed coal	7	4	4	4	450	4
Solid residue	40	20	40	20	240	20
Oil	0.03	0.004	0.004	0.03	0.5	[b]

[a] Mineral matter, 7.3%. High Vol. A. Total S, 3.9% (S_{pyr} 0.13). Ash content of oil, 90 ppm.

mineral matter content of the coal, will be used for gasification or combustion. The leachability of toxic elements should be studied.

One important potential use for low sulfur fuels oils from coal is firing in gas turbines for electricity generation. A particularly clean fuel is needed for this purpose, to avoid damaging the turbine blades. According to F. Robson (2), desired specifications are: ash, 100 ppm; S, < 1.3%; Na + K, < 0.2–0.6 ppm; V, 0.1–0.2 ppm; Pb, 0.1 ppm; Ca, 0.1 ppm; and Cu, 0.02 ppm. It will be seen that the oils analyzed meet many of these specifications. The chief exceptions are the contents of sodium and calcium.

A final point of interest is that in two of the oils, titanium is considerably enriched relative to other elements, compared with the ratios in the original coal. A plausible explanation is that the titanium is present in organometallic combination in these oils.

Acknowledgments

The authors are indebted to the RANN Division of the National Science Foundation for support of their work (Grant No. GI-38974) and

from the B Seam, King Mine, Utah, PSOC-239 (ppm)[a]

Na	Pb	Zn	Mn	Cr	Cu	Cd	Be
5000	[b]	25	12	25	12	0.2	0.4
1000	[b]	50	25	25	25	0.5	0.8
5	0.02	0.12	0.05	0.05	0.02	—	0.0004

[b] Not detected

Lower Dekoven Seam, Will Scarlett Mine, Ill., PSOC-284 (ppm)[a]

Na	Pb	Zn	Mn	Cr	Cu	Cd	Be
2500	[b]	125	125	25	25	4	0.8
8000	[b]	—	300	80	90	4	2
2.5	[b]	0.12	0.12	0.12	0.03	—	0.01

[b] Not detected

from the Mining City Seam, Jerry Mine, Ky., PSOC-221 (ppm)[a]

Na	Pb	Zn	Mn	Cr	Cu	Cd	Be
150	[b]	[b]	4	<1.5	7	1.4	0.2
800	[b]	100	80	20	40	2.7	2
0.9	[b]	0.05	0.05	0.05	0.01	—	0.0009

[b] Not detected

to D. C. Cronauer of Gulf Research and Development Co. for the samples of oils and of hydrogenation residues.

Literature Cited

1. Bishop, M., Ward, D. L., *Fuel* (1958) **37**, 191.
2. Robson, F., personal communication, 1974.

RECEIVED July 22, 1974

15

Trace Element Emissions: Aspects of Environmental Toxicology

ELLIOT PIPERNO

McNeil Laboratories, Inc., Fort Washington, Penn. 19034

This general review of the toxicity of coal-based trace elements emphasizes those which are currently of environmental concern. Increased potential health risks are now associated with those elements which are highly volatilized (e.g., mercury, selenium, and arsenic) and discharged principally as submicron particulates (e.g., lead, cadmium, and nickel). Extensive references to general pharmacologic and toxicologic laboratory data provide a basis for predicting the biological consequences of excessive trace element exposure. Parameters and mechanisms of injury are known, in many instances, however, tolerable body burdens for each of the trace elements must be defined.

The increased reliance on coal to satisfy the nation's growing energy demands superimposed on acute environmental awareness by the public have aroused concern for the pollution consequences. While expansion and environmental quality are not necessarily mutually exclusive, in times of energy crises, air pollution standards commonly yield. The reason is that the consequences of energy shortage, unlike those of air pollution, are known quantifiable entities. Air pollution is associated with many more unknowns, not the least of which are definitions of tolerable burdens.

Interest in trace element emissions received impetus from published reports of widespread atmospheric dissemination of these substances (*1*), especially mercury (*2, 3*), as a consequence of fossil fuel use. More current information indicates that the original mercury discharge estimates were overstated, since they were based on nonrepresentative, ore-associated coal samples (*4*). Of the fossil fuels, coal is considered the major source of atmospheric pollution (*5*) and, compared with oil and uranium use, probably constitutes a greater public health risk (6). Fuel combustion in stationary sources accounts for 21% of total air pollution emissions, and particulates represent 13% of the air pollutants (7).

The paucity of available information on the biological consequences of coal use emissions has been underlined recently by Dixie Lee Ray (8) in her energy report to the President and during a conference sponsored jointly by the American Academy of Pediatrics and the Committee on Environmental Hazards (9). The lag in knowledge of coal-related trace element emissions and their environmental impact has been attributed in large part to belated federal funding (10, 11).

Current experimental work in this area is still by and large in the analytical chemistry phase of development. Pertinent biological studies have not yet made significant headway because they are dependent on a firm chemistry base. Principal sponsors of current research, in addition to the Department of Interior, include the Environmental Protection Agency (EPA), Tennessee Valley Authority (TVA), Atomic Energy Commission (AEC), National Science Foundation (NSF), and utility companies (12).

Trace element reviews which are suggested for further reading and bibliography sources include Refs. 13, 14, 15, and 16 and the complete issue of *Annals of the New York Academy of Science*, Volume 199, "Geochemical Environment in Relation to Health and Disease," (1972).

The Dilemma

No distinct causal relationships have been shown to exist between a pathological entity and coal-related trace element pollution; this of course does not mean there are none. The threat is logically deduced from known relationships arising principally from other industrial sources and statistical findings supporting a relationship between morbidity and/or mortality and coal combustion emissions (6, 17). This forms the basis of the controversy surrounding the magnitude of the threat. Overestimation or excessive concern is often incompatible with economic growth and breeds transient public hysteria and eventual apathy. Underestimation creates conditions for catastrophe because of the large population at risk and the non-uncommon, irreversible consequences that have been associated with industrial contamination accidents.

The goals of protecting public health and establishing a scientifically irrefutable definition of the tolerance limits of trace element exposure are conflicting to a degree since the latter presupposes injury as an endpoint. Therefore, of necessity, errors in setting standards must always be made on the side of overestimating the threat. Since the success of standards enforcement is a function of the soundness of its scientific bases (18), it behooves the scientific community to reduce the error of overestimation.

The trace element threat must be defined in quantitative as well as qualitative terms. It is no easy matter when one considers the number of years it took to establish the relationship between cigarette smoking and lung cancer, where statisticians had the luxury of a control group and groups of graded exposures with which to establish a dose–response relationship. The identity of the carcinogen(s) is still a subject of controversy.

The hazard of coal-related trace element pollution must be considered in conjunction with similar pollution from other sources. There is nothing toxicologically unique about the trace element composition of coal. It is geochemically similar to the makeup of the earth's crust (19), and it contains almost all of the elements on the periodic chart (20). The potential hazard lies in its significant contribution to atmospheric pollution in general, especially in urban areas, and its resultant long term, low level exposures.

The body of experimental toxicologic literature is extremely heterogenous and is largely derived from acute and subacute experimentation. Consequently it does not lend itself easily to meaningful application to the current problem. Safety assessment of long term, low level trace element exposures of necessity requires laboratory experimentation of similar design, at least until reliable short term predictive models can be developed. Short term, high level exposure experiments are gratifying in the sense of producing quick results and perhaps in pinpointing target organs, but their quantitative value and, therefore their predictive usefulness, are questionable. Furthermore, pathological processes dependent upon long latency periods would be missed, such as with carcinogenesis.

Toxicologists have been sensitized by three unique carcinogenic agents in recent history—diethylstilbesterol, because it has caused vaginal adenocarcinoma in young women as late as 25 years after in utero exposure (21); aflatoxin, because it is associated with liver cancer in man after chronic exposure in the ng/kg range (22); and vinyl chloride, because, had it caused a common neoplasm rather than the uncommon hepatic angiosarcoma, the disease process would undoubtedly have been dismissed as "naturally occurring" (23). It was fortunate that these epidemiological discoveries could be reproduced in experimental laboratory studies, otherwise they might have been ignored.

The characteristics of long latency, miniscule insult levels, and nonspecific or low incidence pathology probably confounds trace element research as well. It is well established, for example, that the pulmonary syndrome associated with chronic beryllium exposure (24) and the skin cancer associated with chronic arsenic exposure (25) may have latency periods described in decades. Furthermore, berylliosis only affects a very small percentage of the population exposed.

Historically, trace element toxicity has escaped early diagnosis because it is a relatively rare clinical event; where it was unsuspected, it went undetected. Berylliosis is likely to be misdiagnosed as sarcoidosis (24). Years elapsed before the etiology of Minimata disease was related to mercury and Itai-Itai disease was related to cadmium, even though both conditions were associated with dramatic toxicologic manifestations. Routine clinical laboratory studies, a generally useful diagnostic tool, are often disappointing in trace element toxicity. Such was the case in an outbreak of mercury vapor contamination in a university science department where human urine and laboratory vapor concentrations were 8 and 10 times higher than the acceptable limits, respectively (26).

Establishing causal relationships between trace element exposure and specific pathological processes in the environment is hampered by complex, excessive variables which include populations of varying susceptibility, the concurrent existence and interactions of other stresses (pollutants and nonpollutants), and the normal interfering geological background of trace elements, just to name a few. Understanding the environmental health aspects of trace element pollution is a multidimensional problem requiring a multidisciplinary approach. Neither epidemiological or laboratory data by themselves are sufficient. The meaningful interpretation of the former is frequently plagued by too many variables and the latter by too few.

Fallout and Fate

The composition of trace element emissions during coal combustion is described by Bolton *et al.* in Chapter 13. The actual quantities are somewhat variable depending upon the coal source, the combustion process, the pollution abatement equipment, and the assay itself. Much less is known about important local concentrations of emissions in and around the source, their chemical and physical characteristics, and their fate in the environment.

Solely on the basis of volatility profiles, fossil fuel burning is expected preferentially to transfer As, Hg, Cd, Sn, Sb, Pb, Zn, Tl, Ag, and Bi to the atmosphere (1). In a study designed to detect fallout from a major coal burner equipped with a precipitator, Klein and Russell (27) showed that Ag, Cd, Co, Cr, Fe, Hg, Ni, Ti, and Zn were deposited in the surrounding soil (115 sq mi), and with the exception of mercury, enrichment correlated with the respective metal concentrations in the coal. Mercury was more widely disseminated to the environment. Previous work has indicated that mercury exists primarily in the volatile phase of the flue gas and consequently as much as 90% bypasses the electrostatic precipitation control device (2). Bolton and co-workers have evidence that selenium and arsenic may present a similar problem (*see* Chapter 13).

The similar emission factors reported for mercury and nickel (28), even though the latter is approximately 1000 times more concentrated in coal, probably reflect volatility differentials. Volatile elements commonly considered the most hazardous are: Be, F, As, Se, Cd, Pb, and Hg (4). Volatility and dispersion are associated with multimedia contamination (7) which necessitated the determination of likely absorption through inhalation and ingestion when national ambient standards for mercury were developed (29).

Beryllium and mercury were two of the three air pollutants promulgated as hazardous by the EPA in 1973 and subject to stringent controls (30). While coal combustion releases these elements to the atmosphere, EPA, on the basis of available information, held that this source did not generate sufficient concentrations to be considered hazardous even under restrictive dispersion conditions (29, 31). Coal combustion is thought to be a significant industrial source of atmospheric As (13), Bi (14), Cd (32), Fl (33), Hg (3), and Ni (34).

An understanding of the environmental fate of these elements is necessary in the total assessment of associated health risks. Mercury is known to cycle between the geosphere and biosphere (35). Once in the hydrosphere, it can be converted by sediment flora into highly toxic methylmercury whereupon it is incorporated into aquatic life and ultimately accumulates in human food chains (31). Limited bacterial conversion of inorganic to organic mercury has been shown to occur in soil humus (36) and in animal tissue as well (37). There is no evidence that alkylated mercury is generated from coal combustion directly; if it did it would probably be dissociated to the elemental form (14).

On the basis of chemical profile, Wood (38) predicted that arsenic, selenium, and tellurium will be methylated in the environment, and lead, cadmium, and zinc will not. Elemental concentration in the aquatic food chain has been reported for As (39), Hg (40), Cd (41), Pb (42), and Cu (43). The biological half-life of methylmercury in fish, for example, is one to two years (44). Pillay et al. (40) implicated heavy coal burning in the mercurial contamination of plankton and fish populations of Lake Erie. Other metals, notably cadmium, have been shown to be incorporated into the grazing grasses surrounding a coal burning source (27). Trace element contamination, therefore, can enter the food chain at various points. Disposal of solid wastes in the form of ash and slag is yet another environmental consideration (45).

Identifying the physical and chemical form of a trace element pollutant at the time of body penetration is probably the single most important prerequisite for meaningful biological testing. It is known, for example, that elemental arsenic (13) and beryllium (46) are nontoxic to man compared with their salts. Elemental mercury is highly toxic in the vapor form (*35*) but not when ingested orally (*14*).

With respect to coal emissions, mercury, for example, was found to exist principally (as much as 96%) in the elemental form (*30*). Previously, it was argued by some that ultraviolet radiation transformed it to the less toxic mercuric oxide (*30*). Sunlight tends to degrade mercurial compounds to the elemental form (*47*). Beryllium emissions from coal combustion may be in the nontoxic elemental form (*46*), but this is not known for certain. Fluoride, which is generally assumed to be 100% volatized (*19*), may be trapped with lime in particulates (*33*), but this also is questionable. Highly toxic nickel carbonyl (*48*) and arsine (*49*) emissions have not been reported to date, although the former is a distinct possibility (*50*).

Radioactive materials such as uranium, thorium, and radon gas are released to the environment during coal combustion (*51, 52*). The amounts of radioactivity are often in excess of that released by many modern nuclear power plants but, nevertheless, are well below established radiation standards (*52*).

Particulates

It is estimated that even with 99% efficient fly ash removal, approximately 7000 tons of ash are discharged into the air each year from a 3000-MW coal burning station (*53*). Inhalation of trace element-containing particulates is a direct entry route into the body. The fate of these particulates largely depends on their size—depth of pulmonary penetration is inversely proportional to diameter. Particulates larger than 10μ in diameter are completely removed before leaving the nasal passage, and those larger than 2μ rarely reach the alveolar sac (*54*). Schroeder (*14*) reported that approximately 25% of inhaled particles settle in lung tissue in insoluble forms, a similar amount is exhaled, and the remaining 50% is diverted to the pharynx, where it is swallowed. Absorptive efficiency for most trace elements in the stomach is 5–15% (*55*).

Those submicron particulates which enter the alveolar sacs may undergo various degrees of absorption, depending upon the solubility of their components, or are transported to the base of the ciliated bronchiolar epithelium (*54*). Alveolar absorptive efficiency for most trace elements is 50–80% (*50*). Retention or absorption is not necessarily a simple function of solubility. Silver iodide, for example, is rapidly absorbed from the lungs even though it is weakly soluble in water (*56*). Likewise, insoluble elemental lead deposited in the respiratory passages is absorbed, but the mechanism involved remains to be elucidated (*49*). Vanadium probably accumulates in human lungs in insoluble forms

(57) and is a pulmonary irritant at high concentrations ($\geq 50 \ \mu g/M^3$) (58). Increased ambient particulate load (54) or metallic inhibitors of ciliary activity (14) promote retention and accumulation.

In an analysis of airborne coal fly ash, Natusch and co-workers (50) found that 12 elements, *i.e.*, Pb, Tl, Sb, Cd, Se, Zn, As, Ni, Cr, S, Be, and Mn, were concentrated in the smallest diameter particles. Mercury, although not studied, was expected to follow suit because of its high volatility and probable deposition on small particles. Toca and Berry reported similar findings for lead and cadmium (5). Atmospheric vanadium (59, 60) as well as selenium, antimony, and zinc (61) arising principally from residual fuel combustion also showed a similar pattern. The health risk of this concentration phenomenon is enhanced because of the magnitude of fine particulate emissions and the ease with which these particles bypass particle collection devices, resist fallout, and readily disseminate (50).

Stable aerosols of fine particulates as well as vapors constitute the greatest health risk because of the likelihood of pulmonary absorption. Correlations between trace element pollution and their concentrations in biological fluids or tissue are not uncommon and have been documented for arsenic (62) and lead (63). Man can absorb 75–85% of inhaled mercury vapor at concentrations of 50–350 $\mu g/M^3$ (64) and even more at lower concentrations (65). Certain aerosols like vanadium, iron, manganese, and lead may contribute to the formation of secondary atmospheric pollutants (52, 66).

As previously stated, the magnitude of the trace element pollution threat is a subject of controversy. Schroeder (14) considers atmospheric cadmium, lead, nickel, and mercury to be major public health hazards. Lee and von Lehmden (59) include vanadium because of its predominantly submicron aerosol form. Woolrich (67) contends that other than occupational exposure to Be, Sb, Bi, Sn, Cd, Pb, and Hg, some trace elements in the air are not currently threatening public health, and this is in general agreement with the views of Stern (68). Questioning ambient standards, Stern points out that in some cases, natural background emissions exceed national ambient standards. Jepsen (69) reports that natural mercury emissions in New Almaden, Calif. yielded levels as high as 1.5 mg/M^3 while the national standard is 0.1 mg/M^3. Yet prolonged exposure to 100 $\mu g/M^3$ of mercury is considered by some to be a distinct risk of toxicity (70).

General Toxicology of Trace Elements

Biochemical Aspects. The chemical and physical dissimilarities of trace elements account for their wide scope of toxicologic manifestations.

Many heavy metals, however, share common properties which may serve as a basis for their toxicity. Mercury, cadmium, and lead, for example, all inhibit a large number of enzymes having functional –SH groups, bind to and affect the conformation of nucleic acids, and disrupt pathways of oxidative phosphorylation (*16*). The organ systems having the greatest metabolic requirements (*e.g.*, gastrointestinal mucosa and blood forming tissue) or the lowest reserve capacity to carry on their overall function (*e.g.*, renal tubules and nervous system) tend to show early signs of toxicity (*54*).

Heavy metals stimulate or inhibit a wide variety of enzyme systems (*16, 71, 72*), sometimes for protracted periods (*71, 73*). These effects may be so sensitive as to precede overt toxicity as in the case of lead-induced inhibition of δ ALA dehydrase activity with consequential interference of heme and porphyrin synthesis (*15, 16*). Urinary excretion of δ ALA is also a sensitive indicator of lead absorption (*74*). Another erythrocytic enzyme, glucose-6-phosphatase, when present in abnormally low amounts, may increase susceptibility to lead intoxication (*75*), and for this reason, screens to detect such affected persons in lead-related injuries have been suggested (*76*). Biochemical bases for trace element toxicity have been described for the heavy metals (*16*), selenium (*77*), fluoride (*78*), and cobalt (*79*). Heavy metal metabolic injury, in addition to producing primary toxicity, can adversely alter drug detoxification mechanisms (*80, 81*), with possible secondary consequences for that portion of the population on medication.

From a toxicologic viewpoint, the classification of trace elements into essential and nonessential has probably been overemphasized since both categories produce toxicity. Excesses of one trace element can produce toxicity by decreasing the availability or utilization of another. Thus, excess molybdenum aggravates copper deficiency anemia in ruminants (*82*), excess zinc induces copper deficiency in rats (*83*), excess calcium induces zinc deficiency in pigs (*84*) and increases the manganese requirement of growing chicks (*85*), and excess cadmium induces negative calcium balance in man and copper deficiency in the rat (*86*).

The biochemical basis of cationic antagonisms are largely unknown and often complex. Zinc, the normal cofactor of the metallocarboxypeptidase enzyme can be displaced by mercury, cadmium, or lead causing the enzyme's peptidase, but not its esterase function, to be lost (*16*). Cation–anion interferences in biological systems include cobalt-induced iodine deficiency in children (*87*) and fluoride-induced hypocalcemia in man (*78*). Antagonisms also can be therapeutically exploited. Zinc, iron, manganese, and selenium protect against cadmium toxicity (*88*), molybdenum protects against copper toxicity (*89*), aluminum protects against fluoride toxicity (89), selenium protects against mercury toxicity (90), and mercury, copper, and cadmium protect against selenium toxicity (91).

Many of the trace elements have been used with therapeutic intent. The addition of fluoride to drinking water is the best known example from the standpoint of wide-scale population exposure. Fluoride content in water has been associated with decreased incidences of osteoporosis in women and aortic calcification in men (92). The margin of safety (toxic dose/physiologic dose) for the essential trace elements are generally regarded to be wide. Hoekstra (93) reported safety margins in the range of 40–200, with a typical value of 50. The safety margin of selenium in man is suspected to be low (94) although in laboratory animals it is approximately 100 (4000 ppb toxic dose/40 ppb physiologic dose) (95).

The factors affecting trace element toxicity are bound by the pharmacological principles affecting the toxicity of any chemical and include the physical and chemical characteristics of the substance, its dose or concentration, exposure time, absorption route, organ distribution, metabolic fate, and the organism's defense mechanisms in general. Heavy metals pose a significant threat because they tend to have long biological half-lives, and consequently the opportunity for injury is increased. Lead, for example, has a total body half-life of approximately two months in man (54), and its half-life in bone is estimated at 11 years (96). Cadmium is estimated to have a half-life of 10–25 years in man (97). In addition to lead (98) and cadmium (99), arsenic (49) and mercury (70) also are known to accumulate in man.

Increased susceptibility of certain segments of the population, e.g., young children, may arise from increased tissue sensitivity, more complete absorption, altered distribution, or less developed or impaired defense mechanisms. The increased sensitivity of the child to lead toxicity is well documented (100, 101). In children, unlike the adult, renal tubular damage and encephalopathy are more common sequelae (76, 100). With arsenic exposure, children show significantly higher concentrations of the element in hair and urine than do adults (62).

Tolerance to heavy metals, specifically mercury and cadmium, has been associated with the induction of kidney metallothionein, a protein rich in sulfhydryl groups which protects by chelation (102). The synthetic antidote dimercaprol, introduced after World War I for arsenic-containing gases, works by a similar mechanism (103).

General Somatic Injury. The literature is replete with descriptions of trace element-induced organ system injury. The most recent comprehensive review on this subject is by Louria, et al. (13).

Carcinogenicity. Neoplasia in man has been implicated as a consequence of arsenic exposure both by the oral (skin cancer) (104) and inhalation routes (lung cancer) (105) with the latter association being more controversial. Occupational exposure to airborne beryllium (liver, lung, bile duct, and gallbladder neoplasia) (*24*), chromates (lung cancer) (*105*), and nickel (cancer of the lung and paranasal sinuses) (*105*) have also been implicated in man. In experimental animals, beryllium (*100*), cadmium (*107*), nickel (*48*), and selenium (*100*) have been carcinogenic; arsenic (*49*) and vanadium (*109*) have not. A possible antineoplastic effect has been reported in experimental studies with arsenic (*49*) and in both experimental and epidemiological studies with selenium (*110, 111*).

Genetic and Neonatal Toxicity. The ability of heavy metals readily to cross the placenta and disrupt nucleic acids coupled with the high sensitivity of the fetus and neonate increases the potential dangers of congenital and neonatal toxicity. In mammalian leukocyte cultures, chromosomal aberrations have been reported with lead (*112*), arsenic (*113*), mercury (*114*), and methylmercury (*115*). Charbonneau, *et al.* (*116*) reported a lack of mutagenic effect for methylmercury.

Teratogenicity in experimental animals has been demonstrated with arsenic (*117*), cadmium (*118*), lead (*119*), and inorganic (*114*) and organic mercury (*120*). The organic form of mercury crosses the placenta more readily than the inorganic form (*121*). In combination, lead and cadmium can exert either synergistic or antagonistic effects (*118*). Zinc exerts a protective effect on cadmium-induced malformations even though it does not block the placental transfer of the latter (*122*). Cadmium-induced teratogenicity may be a consequence of altered cadmium:zinc ratios (*122*). These elements are likely to be isomorphic and may compete for the same binding sites on enzymes (*16*). Selenium protects against arsenic or cadmium-induced teratogenicity (*117*).

In addition to teratogenicity, mercury has been associated with decreased reproductive performance in quail (*123*) and lead with neuropathological changes in the suckling rat (*124*). Both inorganic and organic mercury are excreted into the milk with equal ease (*121*). Fluoride is known to be excreted by the lactating breast as well (*78*).

Industrial exposure of pregnant women to lead has resulted in increased incidences of abortion and stillbirths and congenital neurological (*101, 125*) damage. Accidental exposure of pregnant women to methylmercury with congenital neurological sequelae is well documented (*126*). During this time the mother may remain asymptomatic (*127*). In one report, selenium was implicated as a possible human teratogen (*128*).

Immunological Injury. Defects in immunological function may be manifested by such conditions as recurrent infection, autoimmune disease, and lymphoma (*129*). It is hypothesized that heavy metals may alter the function of sulfur-rich immunoproteins by displacing trace element cofactors (*96*). Lead burdens in mice, rats, and baboons similar to those found in humans were associated with decreased resistance to infection and decreased life span (130). Repeated subclinical doses of lead in mice and cadmium in rats followed by a challenging dose of *Salmonella typhimurium* showed increased mortality, suggesting an impairment of the immune response (131). Lead and cadmium are known to impair the oxidative metabolism of phagocytes and the elaboration of antibody in a variety of animals including mice and rats (129). With viral infections, metal-induced impairment of the immune response may arise from decreased interferon production. Arsenic and lead, but not cadmium, mercury, or nickel, inhibited the protective effect of the interferon-inducing agent, poly I:poly C on encephalomyocarditis virus infection of mice (132).

In studies on rabbit alveolar macrophage cultures, Waters and co-workers (133) presented data suggesting that vanadium oxides may adversely affect pulmonary defense. The cytotoxicity of the oxides studied were directly related to their solubility, *i.e.*, $V_2O_5 > V_2O_3 > VO_2$. Ambient vanadium concentrations in urban regions have been reported to correlate with mortality incidence from bronchitis and pneumonia, especially in males (134). Likewise, industrial exposure to airborne manganese has been shown to correlate with increased incidence of bronchitis, caused in part by increased susceptibility to infection (135).

Impairment of the immune response has also been demonstrated with beryllium in experimental animals (136). The long latency period associated with berylliosis development may in fact be a delayed immune response. This hypothesis is supported by the finding that lymphocyte cultures obtained from chronically exposed workers underwent a high degree of blast formation when challenged with beryllium sulfate (137). The authors suggest that this test may serve as a screen for detecting hypersensitive individuals.

Phytotoxicity and Related Problems. It is difficult if not impossible to generalize on this subject because, in addition to the normal variables governing a plant's response to chemical exposure (*e.g.* soil pH, temperature, humidity), there is the complication of differential species sensitivity. Ragweed grows luxuriantly in high soil concentrations of zinc while the surrounding vegetation is stunted (138); radishes are significantly more sensitive to the toxic effects of cadmium than is lettuce (139); and corn is sensitive to soil iodine concentrations of 8 ppm while lettuce tolerates 160 ppm (138), to cite a few examples. Absorption of trace elements not only varies with species and variety but with plant part as well (140).

Plants represent man's single most important food source and are the main source of trace elements for the general population. When industrial contamination of soil occurs, pollutants may be incorporated in surrounding food crops. Purves (*141*) reported more than twice as much boron, five times as much copper, 17 times as much lead, and 18 times as much zinc in urban compared with rural soils.

Toxicity from the ingestion of contaminated vegetation is a more common occurrence in grazing animals than man because of the limited food source of the former. Fluoride fallout onto grazing grasses from aluminum reduction plants has resulted in dental fluorosis and decreased milk production in cows (*33, 89*). Naturally high soil and plant levels of molybdenum or selenium have resulted in molybdenosis (manifested by copper deficiency) (*138*) and "blind staggers" in grazing animals (*142*), respectively, and yet plant toxicity with these elements is uncommon (*138*).

There are instances, for example with aluminum and copper, where even phytotoxic concentrations pose no health hazard to animals (*143*). Lead contamination, on the other hand, is considered a bigger hazard to man and animal by accidental ingestion than to plants because it is largely unavailable (insoluble) to the latter (*100*). Fortunately, the element is largely removed by simple rinsing (*144*). Fluoride, in contrast to lead, is absorbed readily in the free form and tends to be phytotoxic at extremely low concentrations (*145*).

Human blood levels of selenium are reported to correlate somewhat with soil concentrations of the element (*94*). In addition, human ingestion of selenized vegetables and grain has resulted in signs of selenium toxicity (*146*). Ingestion of rice contaminated with cadmium from the effluent of a mining operation was hypothesized as the etiological factor in Itai-Itai disease in man (*32*).

Heavy metal toxicity in plants is infrequent (*143*). In many cases, metal concentrations in plant parts show poor correlation with soil concentrations of the element (*147*). Plants tend to exclude certain elements and readily accept or concentrate others. Lisk (*148*) reported natural plant:soil concentration ratios of 0.05 or less for As, Be, Cr, Ga, Hg, Ni, and V. Cadmium appears to be actively concentrated and selenium appears to be easily exchangeable. Indicator plants are capable of markedly concentrating specific elements, *e.g.*, *Astragalus spp.* for selenium (*138*) and *Hybanthus floribundus* for nickel (*149*). Plants growing on mine wastes have been shown to evolve populations which exhibit metal-specific tolerances (*150*).

Plant intolerance to metals may be associated with absorption in root cells (*151*). Chromium concentrates in root cells, and that is where it exerts its toxicity (*138*). Intolerance to selenium in certain crop plants high in sulfur, such as cabbage, may be related to an indiscriminate sulfur–selenium substitution in enzyme systems (*77*).

Cationic antagonisms induced by trace elements excesses have been demonstrated in plants as well as animals. Excesses of either manganese or iron in soil have resulted in deficiencies of the other in soybeans (152). Likewise, excess soil concentration of copper has resulted in manganese deficiency in crops (153). Plants, like animals, also demonstrate chromosomal aberrations (particularly to aluminum and cadmium) (15) as well as enzymatic alterations in response to excess exposure (16).

The Future

Hopefully, within the next decade, air pollution toxicology will make significant advances from its current descriptive phase to one of concept development. Descriptive data will always be necessary, but for it to be meaningful, it must be put into perspective. The physical and chemical characterization of trace element emissions and determinations of their environmental fate are prerequisites for meaningful biological testing. Studies on the biological effects of trace element loads in combination as well as individually are needed because of their combined presence in the environment and the complexity of their interactions.

Tolerable trace element burdens and the affecting variables must be appreciated prior to being able to predict a hazard. To this end, the publication of negative results or the absence of detrimental effects should be encouraged along with positive data. Predictive models of toxicity must be sought, and biological monitoring in addition to environmental surveillance should be instituted as a measure of absorption and as an early warning system. Tissue banks of autopsy material have already been suggested (154). Unfortunately, nonspecific alterations of trace element concentrations in biological tissue occur in a wide variety of disease states (155) and will serve as confounding factors. In experimental studies, not only should the final toxic or pathological insult be characterized but pathogenesis (all the preceding events) as well. Such data would be useful in diagnosing subclinical effects. In the meantime, technological advances to reduce trace element emissions should be encouraged vigorously.

Acknowledgments

I gratefully acknowledge the cooperation given by: Elizabeth M. Rountree, Jean M. Ellis, and Emily A. Fisher, of McNeil Laboratories, Inc. for their library services; Robert M. Kirsch and Robert K. Dix of the same organization for their help in collating data and offering suggestions; and Frank Leone of the AEC as well as various members of the EPA for providing the necessary background data and pertinent bibliographies.

Literature Cited

1. Bertine, K. K., Goldberg, E. D., *Science* (1971) **173**, 233.
2. Billings, C. E., Matson, W. R., *Science* (1972) **176**, 1233.
3. Joensuu, O. I., *Science* (1971) **172**, 1027.
4. Hall, H. J., Varga, G. M., Magee, E. M., "Trace Elements and Potential Pollutant Effects in Fossil Fuels," EPA Contract Report **R2-73-249**, PB 225, 039, June 1973.
5. Toca, F. M., Berry, C. M., *Amer. Ind. Hyg. Ass. J.* (1973) **34** (9), 396.
6. Lave, L. B., Freeburg, L. C., *Nucl. Safety* (1973) **14** (5), 409.
7. Newill, V. A., *Pediatrics* (supplement) (1974) **53** (5), Part II, 785.
8. Ray, D. L., "The Nation's Energy Future, A Report to Richard M. Nixon, President of the United States," U. S. Government Printing Office Stock Number **5210-00363**, December 1973.
9. Kotin, P., *Pediatrics* (supplement) (1974) **53** (5), Part II, 782.
10. Rose, D. J., *Science* (1974) **184**, 351.
11. Walsh, J., *Science* (1974) **184**, 336.
12. "Inventory of Current Energy Research and Development," Volumes I, II, and III, U. S. Government Printing Office Stock Numbers **5270-02173** and **5270-02174**, January 1974.
13. Louria, D. B., Joselow, M. M., Browder, A. A., *Ann. Intern. Med.* (1972) **76** (2), 307.
14. Schroeder, H. A., *Environment* (1971) **13** (8), 18.
15. Schubert, J., "Heavy Metals: Toxicity and Environmental Pollution," in "Metal Ions in Biological Systems," S. K. Dhar, Ed.), p. 239, Plenum, New York, 1973.
16. Vallee, B. L., Ulmer, D. D., *Annu. Rev. Biochem.* (1972) **41**, 91.
17. Cohen, A. A., Bromberg, S., Buechley, R. W., Heiderscheit, L. T., Shy, C. M., *Amer. J. Pub. Health* (1972) **62**, 1181.
18. Train, R. E., *Science* (1974) **184**, 1050.
19. Ruch, R. R., Gluskoter, H. J., Shimp, N. F., "Occurrence and Distribution of Potentially Volatile Trace Elements in Coal: An Interim Report," *Ill. State Geol. Surv., Environ. Geol. Notes*, (61), April 1973.
20. Abernethy, R. F., Peterson, M. J., Gibson, F. H., "Spectrochemical Analyses of Coal Ash for Trace Elements," *Bur. Mines (U.S.) Rep. Invest.* **7281**, 1969.
21. Herbst, A. L., Ulfelder, H., Poskanzer, D. C., *New Eng. J. Med.* (1971) **284**, 878.
22. Peers, F. G., Linsell, C. A., *Brit. J. Cancer* (1973) **27**, 473.
23. *Brit. Med. J.* (1974) **1**, 590.
24. *Med. World News* (1973) **14** (39), 32.
25. Fierz, V., *Dermatologica* (1965) **131**, 41.
26. Rose, K. D., Simpson, E. W., Week, D., *J. Amer. Coll. Health Ass.* (1972) **20** (3), 197.
27. Klein, D. H., Russel, P., *Environ. Sci. Tech.* (1973) **7** (4), 357.
28. Anderson, D., "Emission Factors for Trace Substances," EPA Document **450/2-73-001**, December 1973.
29. "Background Information on Development of National Emission Standards for Hazardous Air Pollutants: Asbestos, Beryllium, and Mercury," EPA Publication No. **APTD-1503**, March 1973.
30. *Fed. Regist.* (April 6, 1973) **38** (66), 8820.
31. "Background Information—Proposed National Emission Standards for Hazardous Air Pollutants: Asbestos, Beryllium, and Mercury," EPA Publication No. **APTD-0753**, December 1971.
32. Cox, D. B., *J. Environ. Health* (1974) **36** (4), 361.
33. Prival, M. J., Fisher, F., *Environment* (1973) **15** (1), 25.

34. Bowen, H. J. M., "Trace Elements in Biochemistry," Academic, New York, 1966.
35. Peakall, D. B., Lovett, R. J., *Bioscience* (1972) **22** (1), 20.
36. Wood, J. M., Kennedy, F. C., Rosen, C. G., *Nature* (1968) **220**, 173.
37. Westöö, G., *Acta Chem. Scand.* (1968) **22**, 2277.
38. Wood, J. M., *Science* (1974) **183**, 1049.
39. Jernelöv, A., unpublished data.
40. Pillay, K. K. S., Thomas Jr., C. C., Sondel, J. A., Hyche, C. M., *Environ. Res.* (1972) **5**, 172.
41. Schroeder, H. A., *J. Chronic Dis.* (1967) **20**, 179.
42. Novick, R. E., *J. Environ. Health* (1973) **35** (4), 363.
43. Schroeder, H. A., Nason, A. P., Tipton, I. H., *J. Chronic Dis.* (1966) **19**, 1007.
44. Wood, J. M., *Environment* (1972) **14** (1), 33.
45. Mills, G. A., Johnson, H. R., Perry, H., *Environ. Sci. Technol.* (1971) **5** (1), 30.
46. *Environ. Sci. Technol.* (1971) **7** (5), 584.
47. Johasson, I. R., "Mercury in the Natural Environment, A Review of Recent Work," *Geol. Surv. Can.* (1971).
48. Sunderman, F. W., Donnelly, A. J., *Amer. J. Path.* (1965) **46**, 1027.
49. Harvey, S. C., "Heavy Metals" in "The Pharmacological Basis of Therapeutics," (L. S. Goodman and A. Gilman, Eds.), 4th ed., MacMillan, New York, 1970.
50. Natusch, D. F. S., Wallace, J. R., Evans, C. A., *Science* (1974) **183**, 202.
51. Eisenbud, M., Petrow, H. G., *Science* (1964) **144**, 288.
52. Pennsylvania Dept. of Education, "Environmental Impact of Electrical Power Generation: Nuclear and Fossil," prepared under contract AT (40-1)-4167, U. S. Atomic Energy Commission, 1973.
53. "Electric Power and the Environment," sponsored by the Energy Policy Staff, Office of Science and Technology," Washington, D. C., August 1970.
54. Levine, R. R., "Pharmacology: Drug Actions and Reactions," Little, Brown, and Co., Boston, 1973.
55. Patterson, C. C., *Arch. Environ. Health* (1965) **11**, 344.
56. Willard, D. H., Bair, W. J., *Acta Radiol.* (1961) **55**, 70.
57. Schroeder, H. A., "Vanadium," Air Quality Monograph No. **70-13**, American Petroleum Institute, Washington, D.C., 1970.
58. Waters, M. D., Gardner, D. E., Coffin, D. L., *Environ. Health Perspect.* (June 1973) Experimental Issue (4), 99.
59. Lee, Jr., R. E., von Lehmden, D. J., *J. Air Poll. Control Ass.* (1973) **23** (10), 853.
60. Sugimae, A., Hasegawa, T., *Environ. Sci. Technol.* (1973) **7** (5), 444.
61. Gladney, E. S., Zoller, W. H., Jones, A. G., Gordon, G. E., *Environ. Sci. Technol.* (1974) **8** (6), 551.
62. Milham, Jr., S., Strong, T., *Environ. Res.* (1974) **7** (2), 176.
63. Needleman, H. L., Shapiro, I. M., "Dental Lead Levels in Asymptomatic Philadelphia School Children: Subclinical Exposures in High and Low Risk Groups," EPA-NIEHS Conference on Low Level Lead Toxicity, Raleigh, North Carolina, October 1-2, 1973.
64. *Arch. Environ. Health* (1969) **19**, 891.
65. Magoes, K., *Environ. Res.* (1967) **1**, 323.
66. Urone, P., Lutsep, H., Noyes, C. M., Parcher, J. F., *Environ. Sci. Technol.* (1968) **2** (8), 611.
67. Woolrich, P. F., *Amer. Ind. Hyg. Ass. J.* (1973) **34** (5), 217.
68. Stern, A. C., *Environ. Sci. Technol.* (1974) **8** (1), 16.
69. Jepsen, A. F., "Abstracts of Papers," National Meeting American Chemical Society, Abstract No. **76**, September 1971.

70. Friberg, L., Vostal, J., "Mercury in the Environment—A Toxicological and Epidemiological Appraisal," Karolinska Institute Department of Environmental Hygiene (Stockholm) for the U.S. Environmental Protection Agency (Office of Air Programs), November 1971.
71. Singhal, R. L., Merali, Z., Kacew, S., Sutherland, D. J. B., *Science* (1974) **183**, 1094.
72. Dietar, M. P., *Toxicol. Appl. Pharmacol.* (1974) **27**, 86.
73. Prerovska, I., Teisinger, J., *Brit. J. Med.* (1970) **27**, 352.
74. Haeger-Aronson, B., *Scand. J. Clin. Lab. Invest.* (1960) supplement **12**, 47.
75. McIntire, M. S., Angle, C. R., *Science* (1972) **177**, 520.
76. Kehoe, R. A., *J. Roy. Inst. Pub. Health Hyg.* (1961) **24**, 81.
77. Stadtman, T. C., *Science* (1974) **183**, 915.
78. Welt, L. G., Blythe, W. B., "Anions: Phosphate, Iodide, Fluoride, and Other Anions," in "The Pharmacological Basis of Therapeutics," (L. S. Goodman and A. Gilman, Eds.), 4th ed., Macmillan, New York, 1970.
79. Eriksen, L., Eriksen, N., Haavaldsen, S., *Acta Physiol. Scand.* (1961) **53**, 300.
80. Hadley, W. M., Miya, T. S., Bousquet, W., *Toxicol. Appl. Pharmacol.* (1974) **28**, 284.
81. Fouts, J. R., Pohl, R. J., *J. Pharmacol. Exp. Ther.* (1971) **179**, 91.
82. Dowdy, R. P., Matrone, G., *J. Nutr.* (1968) **95**, 191.
83. Smith, S., Larson, E., *J. Biol. Chem.* (1946) **163**, 29.
84. O'Dell, B. L., *Amer. J. Clin. Nutr.* (1969) **22**, 1315.
85. O'Dell, B. L., *Ann. N.Y. Acad. Sci.* (1972) **199**, 70.
86. *Environ. Sci. Technol.* (1973) **7** (8), 684.
87. Washburn, T. C., Kaplan, E., *Clin. Pediat.* (Phila) (1964) **3**, 89.
88. Pakalne, D., Nollendorf, A. F., Upitas, V., *Latv. PSR Zinat. Akad. Vestis* (1970) **11**, 16.
89. Oehme, F. W., *DVM* (1974) **6** (3), 2.
90. Potter, S. D., Matrone, G., *Environ. Health Perspect.* (June 1973), Experimental Issue (4), 100.
91. Hill, C. H., *Environ. Health Perspect.* (June 1973) Experimental Issue (4), 104.
92. Bernstein, D. S., Sadowsky, N., Hegsted, D. M., Guri, C. D., Stare, F. J., *J. Amer. Med. Ass.* (1966) **198**, 499.
93. Hoekstra, W. G., *Ann. N.Y. Acad. Sci.* (1972) **199**, 182.
94. Mertz, W., *Ann. N.Y. Acad. Sci.* (1972) **199**, 191.
95. Lakin, H. W., "Abstracts of Papers," National Meeting, American Chemical Society, Abstract No. **77**, September 1971.
96. Chisolm, J. J., *Pediatrics* (supplement) (1974) **53** (5), Part II, 841.
97. *Environ. Sci. Technol.* (1971) **5** (9), 754.
98. Emmerson, B. T., *Ann. Intern. Med.* (1968) **68** (2), 488.
99. Friberg, L., Piscator, M., Nordberg, G., "Symposium on Cadmium in the Environment. A Toxicological and Epidemiological Appraisal," University of Rochester, Rochester, N. Y., June 14–18, 1971.
100. Position Paper on the Health Implications of Airborne Lead, U. S. Environmental Protection Agency, Washington, D.C., November 28, 1973.
101. Finberg, L., *Pediatrics* (supplement) (1974) **53** (5), Part II, 831.
102. Piotrowski, J. K., Trojanowska, B., Wisniewska-Knypl, J. M., Bolanowska, W., *Toxicol. Appl. Pharmacol.* (1974) **27**, 11.
103. Oehme, F. W., *Clin. Toxicol.* (1972) **5** (2), 215.
104. Tseng, W. P., Chu, H. M., How, S. W., Fong, J. M., Lin, C. S., Yeh, S., *J. Nat. Cancer Inst.* (1968) **40** (3), 453.
105. Fraumeni, J. F., *Pediatrics* (supplement) (1974) **53** (5), Part II, 807.
106. Barnes, J. M., Denz, F. A., *Brit. J. Cancer* (1950) **4**, 212.
107. Furst, A., Haro, R. T., *Progr. Exp. Tumor Res.* (1969) **12**, 102.

108. Schroeder, H. A., Mitchener, M., *J. Nutr.* (1971) **101**, 1531.
109. Schroeder, H. A., Mitchener, M., Nason, A. P., *J. Nutr.* (1970) **100**, 59.
110. *Med. Trib.* (1973) **14** (24), 27.
111. Shamburger, R. S., Frost, D. V., *Can. Med. Ass. J.* (1969) **100**, 682.
112. Muro, L. A., Goyer, R. A., *Arch. Pathol.* (1969) **87**, 660.
113. Oppenheim, J. J., Fishbein, W. N., *Cancer Res.* (1965) **25**, 980.
114. Malling, H. V., Wasson, J. S., Epstein, S. S., *Newslett. Environ. Mutagen Soc.* (1970) **3**, 7.
115. Skerfving, K., Hansson, K., Lindsten, J., *Arch. Environ. Health* (1970) **21**, 133.
116. Charbonneau, S. M., Munro, I. C., Nera, E. A., *Toxicol. Appl. Pharmacol.* (1974) **27**, 569.
117. Holmberg, R. E., Ferm, V. H., *Arch. Environ. Health* (1969) **18**, 873.
118. Ferm, V. H., *Experientia* (1969) **25**, 56.
119. Ferm, V. H., Ferm, D. W., *Life Sci.* (1971) **10**, 35.
120. Spyker, J. M., Smithberg, M., *Teratology* (1972) **5**, 181.
121. Mansour, M. M., Dyer, N. C., Hoffman, L. H., Schulert, A. R., Brill, A. B., *Environ. Res.* (1973) **6**, 479.
122. Ferm, V. H., Hanlon, D. P., Urban, J., *J. Embryol. Exp. Morphol.* (1969) **22**, 107.
123. Thaxton, P., Parkhurst, C. R., *Environ. Health Perspect.* (June 1973) Experimental Issue (4), 104.
124. Pentschew, A., Garro, F., *Acta Neuropathol.* (1966) **6**, 266.
125. Lee, D. H. K., "Metallic Contaminants and Human Health," p. 117, Academic, New York, 1972.
126. Matsumoto, H., Koya, G., Takeuchi, T., *J. Neuropathol. Exp. Neurol.* (1965) **24**, 563.
127. Snyder, R. D., *New Eng. J. Med.* (1971) **284**, 1014.
128. Robertson, D. S. E., *Lancet* (1970) **1**, 518.
129. Bellanti, J. A., *Pediatrics* (supplement) (1974) **53** (5), Part II, 818.
130. Adamson, L., *Environment* (1973) **15** (1), 21.
131. Hemphill, F. E., Kaeberle, M. L., Buck, W. B., *Science* (1971) **172**, 1031.
132. Gainer, J. H., *Environ. Health Perspect.* (June 1973) Experimental Issue (4), 98.
133. Waters, M. D., Gardner, D. E., Coffin, D. L., *Toxicol. Appl. Pharmacol.* (1974) **28**, 253.
134. Stocks, P., *Cancer* (1960) **14**, 397.
135. Davies, T. A., *Brit. J. Ind. Med.* (1946) **3**, 11.
136. Vacher, J., Deraedt, R., Benzoni, J., *Toxicol. Appl. Pharmacol.* (1974) **28**, 28.
137. Deodar, S. D., Barna, B., Van Ordstrand, H. S., *Chest* (1973) **63** (3), 309.
138. Hemphill, D. D., *Ann. N.Y. Acad. Sci.* (1972) **199**, 46.
139. John, M. K., Van Laerhoven, C. J., Chuah, H. C., *Environ. Sci. Technol.* (1972) **6** (12), 1005.
140. Kirkham, M. B., *Science* (1974) **184**, 1030.
141. Purves, D., *Environ. Poll.* (1972) **3** (1), 17.
142. Kubota, J., Allaway, W. H., Carter, D. L., Carey, E. E., Lazar, V. A., *Agr. Food Chem.* (1967) **15**, 488.
143. Horvath, D. J., *Ann. N.Y. Acad. Sci.* (1972) **199**, 82.
144. Schuck, E. A., Locke, J. K., *Environ. Sci. Technol.* (1970) **4** (4), 324.
145. Wood, F. A., *Phytopathology* (1968) **58** (8), 1075.
146. Lemley, R. E., Merryman, M. P., *Lancet* (1941) **61**, 435.
147. Shacklette, H. T., Sauer, H. I., Miesch, A. T., *Geol. Soc. Amer. Bull.* (1972) **83** (4), 1077.
148. Lisk, D. J., *Advan. Agron.* (1972) **24**, 267.
149. Severne, B. C., *Nature* (1974) **248**, 807.

150. Antonovics, J., *Environ. Health Perspect.* (June 1973) Experimental Issue (4), 103.
151. Smith, R. A. H., Bradshaw, A. D., *Inst. Mining Met. Trans. Sect. A* (1972) **81**, A 230.
152. Somers, I., Shive, J., *Plant Physiology* (1942) **17**, 582.
153. Albert, A., "Selective Toxicity," 4th ed., p. 303, Methuen, London, 1968.
154. Hammer, D. I., Colucci, A. V., Creason, J. P., Pinkerton, C., *Environ. Health Perspect.* (June 1973) Experimental Issue (4), 97.
155. McCall, J. T., Goldstein, N. P., Smith, L. H., *Federation Proc.* (1971) **30** (3), 1011.

RECEIVED August 26, 1974

INDEX

The text of this book is set in 10 point Caledonia with two points of leading. The chapter numerals are set in 30 point Garamond; the chapter titles are set in 18 point Garamond Bold.

The book is printed offset on Danforth 550 Machine Blue White text, 50-pound. The cover is Joanna Book Binding blue linen.

Jacket design by Linda McKnight.
Editing and production by Virginia Orr.

The book was composed by the Mills-Frizell-Evans Co., Baltimore, Md., printed and bound by The Maple Press Co., York, Pa.